John Butler Johnson

Johnson's tables. Stadia and Earthwork Tables,

Four-place logarithms, logarithmic traverse table, natural functions, Map projections

John Butler Johnson

Johnson's tables. Stadia and Earthwork Tables,
Four-place logarithms, logarithmic traverse table, natural functions, Map projections

ISBN/EAN: 9783337024574

Printed in Europe, USA, Canada, Australia, Japan

JOHNSON'S TABLES.

STADIA AND EARTH-WORK TABLES.

FOUR-PLACE LOGARITHMS, LOGARITHMIC TRAVERSE TABLE, NATURAL FUNCTIONS, MAP PROJECTIONS, ETC., ETC.

REPRINTED FROM

THEORY AND PRACTICE OF SURVEYING.

BY

J. B. JOHNSON,

PROFESSOR OF CIVIL ENGINEERING, WASHINGTON UNIVERSITY, ST. LOUIS

NEW YORK:

JOHN WILEY & SONS,

53 EAST TENTH STREET.

1892.

NOTE BY THE AUTHOR.

THE great use made by engineers of three of the following tables, viz., the Four-place Logarithmic Table, the Stadia Table, and the table giving Prismoidal Volumes, has necessitated the binding of these in more convenient form than that in which they first appeared in the *Theory and Practice of Surveying.* Since the cost is not materially increased by additional pages, the remaining tables are also included, as well as the entire chapter on the Measurement of Volumes.

The Stadia Tables were computed by Mr. Arthur Winslow, State Geologist of Missouri, and first published by the Pennsylvania Geological Survey. The four-place logarithm tables were originally taken from Lee's Tables and Formulæ, a publication of the U. S. Engineer Corps. The table giving Volumes by the Prismoidal Formula was computed by the Author. It is the only table, he believes, giving volumes by the prismoidal formula at one operation. It may also be used for Mean End-areas. Tables IV and VIII are also original in their arrangement.

J. B. J.

EXPLANATION OF TABLES.

TABLES I, II, III, VI, and VII require no explanation.

TABLE IV gives logarithmic sines and cosines to four places for computing latitudes and departures when the angles are read from zero to 360 degrees. It can of course be used for bearings reading from zero to 90 degrees, as is ordinarily done in compass work. In stadia work, and always in transit work where the instrument is graduated continuously to 360 degrees, this table will be found very convenient for coördinating traverse lines, as well as for computing latitudes and departures for closed surveys.

From zero to 5 degrees, and from 85 to 90 degrees, the tables give values for each minute of arc without tabular differences. From 5 to 45 degrees values are given for each 10 minutes of arc with tabular differences for the log. sines, and from 45 to 85 degrees with tabular differences for the 10-minute increments for the log. cosines. In the other cases the tabular difference is so small as to be readily taken at sight. Table III_A can of course be used in place of Table IV if preferred.

TABLE V gives *horizontal distance* and *difference of elevation* for inclined sights in stadia work. The true equations of reduction are:

$$Hor.\ Dist. = r \cos^2 v + (c + f) \cos v, \quad . \quad . \quad . \quad . \quad (1)$$

and

$$Dif.\ Elev. = r \cos v \sin v + (c + f) \sin v; \quad . \quad . \quad (2)$$

where

$r =$ reading of distance on stadia rod when held vertically;
$v =$ vertical angle with the horizon;
$f =$ focal length of objective;
$c =$ distance from objective to centre of instrument.

The tables give the values for the first term only of the second member. The values for the second term are given at the bottom of the page, the constant term $(c+f)$ in the above equations being there called "c." The sum of these two distances, viz., distance from centre of instrument to objective plus distance from cross-wires to objective, varies in different instruments from nine to fifteen inches. Three values of this second term are given, therefore, one corresponding to $c+f=$ 0.75 foot, one to $c+f=1.00$ foot, and one to $c+f=1.25$ foot. In ordinary work these corrections may be neglected. See chapter on Stadia Surveying in the *Theory and Practice of Surveying*.

A Reduction Diagram, printed from an engraved plate 20 by 24 inches, has been prepared with great care, giving corrections to the horizontal distance read, and the differences of elevation, for inclined sights, as shown by the table, not including the $(c+f)$ term. For all angles below 6° and distances less than 1500 feet, with differences of elevation less than 50 feet, this diagram is much preferable to the table. The results are found at one operation, to the nearest tenth of a foot, with great rapidity. It can be procured from the publisher of these tables, printed on heavy lithographic paper, price 50 cents, post paid.

TABLE VIII gives the coördinates to be used in the polyconic projection of maps. It is fully explained in the chapter on Projection of Maps in the *Surveying*.

TABLES IX and X will be found very useful in sewer and hydraulic work where Kutter's formula is to be used. They are fully explained in the chapter on Hydrographic Surveying.

TABLE XI gives correct volumes of prismoids, by the prismoidal formula.

For the benefit of railroad engineers and others who either do not possess a copy of the *Surveying*, or who do not have it by them, the entire chapter on the Measurement of Volumes is here inserted. At least seven pages of this chapter is requisite to a full explanation of the table, and for the sake of completeness, and to show the superiority of this table over any table of volumes from mean end-areas, or by the use of diagonals, it has been thought best to insert the entire chapter.

TABLE XII gives the azimuth of Polaris at any hour-angle. By its use an observation for azimuth to the nearest minute of arc can be made at any hour when the star is visible, provided the local time is known to within one or two minutes. When the observation is taken two hours from the time of elongation, the local time need not be known nearer than five minutes. A detailed explanation of its use is given in the *Surveying*, Art. 381A.

CONTENTS.

TABLES.

CONTENTS.

CHAPTER XIII.

THE MEASUREMENT OF VOLUMES.

310. Proposition.—*The volume of any doubly-truncated prism or cylinder, bounded by plane ends, is equal to the area of a right section into the length of the element through the centres of gravity of the bases, or it is equal to the area of either base into the altitude of the element joining the centres of gravity of the bases, measured perpendicular to that base.*

Let $ABCD$, Fig. 107, be a cylinder, cut by the planes OC and OB, the unsymmetrical right section EF being shown in plan in $E'F'$. Whatever position the cutting planes may have, if they are not parallel they will intersect in a line. This line of intersection may be taken perpendicular to the paper, and the body would then appear as shown in the figure, the line of intersection of the cutting planes being projected at O.

Let A = area of the right section;
ΔA = any very small portion of this area;
x = distance of any element from O;
then ax = height of any element at a distance x from O.

An elementary volume would then be $ax\Delta A$, and the total volume of the solid would be $\Sigma ax\Delta A$.

Again, the total volume is equal to the mean or average height of all the elementary volumes multiplied by the area of the right section.

The mean height of the elementary volumes is, therefore,

$\frac{\Sigma ax\Delta A}{A} = \frac{a\Sigma x\Delta A}{A}$. But $\frac{\Sigma x\Delta A}{A}$ is the distance from O to the centre of gravity, G, of the right section,* and a times this distance is the height of the element LK through this point. Therefore, the mean height is the height through the centre of

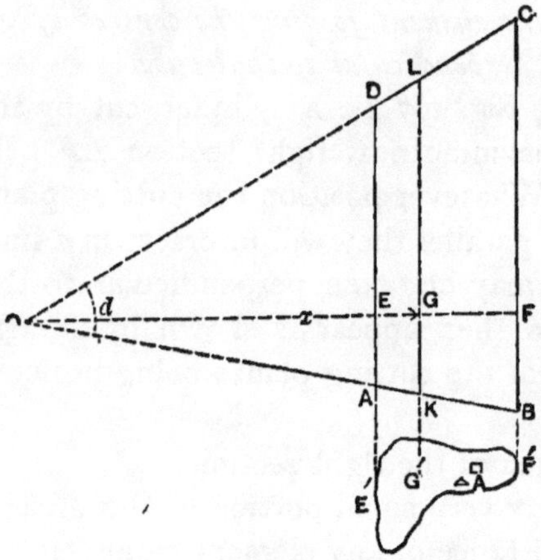

FIG. 107.

gravity of the base, and this into the area of the right section is the volume of the truncated prism or cylinder. The truth of the alternative proposition can now readily be shown.

Corollary. When the cylinder or prism has a symmetrical cross-section, the centre of gravity of the base is at the centre of the figure, and the length of the line joining these centres is the mean of any number of symmetrically chosen exterior elements. For instance, if the right section of the prism be a regular polygon, the height of the centre element is the mean of the length of all the edges. This also holds true for parallelograms, and hence for rectangles. Here the centres of gravity

* This is shown in mechanics, and the student may have to take it for granted temporarily.

of the bases lie at the intersections of the diagonals; and since these bisect each other, the length of the line joining the intersections is the mean of the lengths of the four edges. The same is true of triangular cross-sections.

311. **Grading over Extended Surfaces.**—Lay out the area in equal rectangles of such a size that the surfaces of the several rectangles may be considered planes. For common rolling ground these rectangles should not be over fifty feet on a side. Let Fig. 108 represent such an area. Drive pegs at

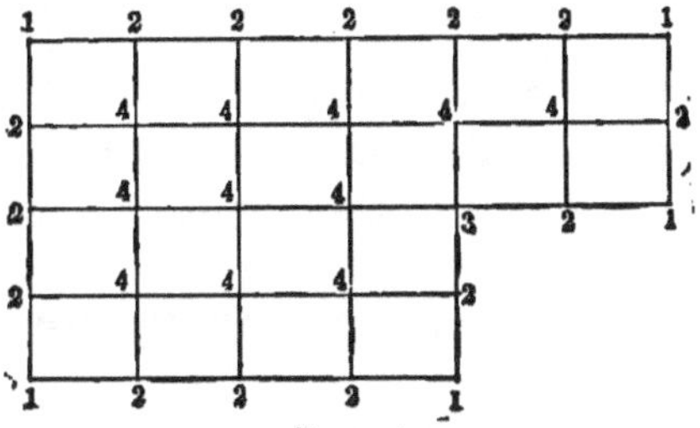

FIG. 108.

the corners, and find the elevation of the ground at each intersection by means of a level, reading to the nearest tenth of a foot, and referring the elevations to some datum-plane below the surface after it is graded. When the grading is completed, relocate the intersections from witness-points that were placed outside the limits of grading, and again find the elevations at these points. The several differences are the depths of excavation (or fill) at the corresponding corners. The contents of any partial volume is the mean of the four corner heights into the area of its cross-section. But since the rectangular areas were made equal, and since each corner height will be used as many times as there are rectangles joining at that corner, we have, in cubic yards,

$$V = \frac{A}{4 \times 27} [\Sigma h_1 + 2\Sigma h_2 + 3\Sigma h_3 + 4\Sigma h_4]. \quad . \quad . \quad (1)$$

The subscripts denote the number of adjoining rectangles the area of each of which is A.

From this equation we may frame a

RULE.—Take each corner height as many times as there are partial areas adjoining it, add them all together, and multiply by one fourth of the area of a single rectangle. This gives the volume in cubic feet. To obtain it in cubic yards, divide by twenty-seven.

If the ground be laid out in rectangles, 30 feet by 36 feet, then $\frac{A}{4 \times 27} = \frac{1080}{108} = 10$; and if the elevations be taken to the nearest tenth of a foot, then the sum of the multiplied corner heights, with the decimal point omitted, is at once the the amount of earthwork in cubic yards. This is a common way of doing this work. In borrow-pits, for which this method is peculiarly fitted, the elementary areas would usually be smaller.

In general, on rolling ground, a plane cannot be passed through the four corner heights. We may, however, pass a plane through any three points, and so with four given points

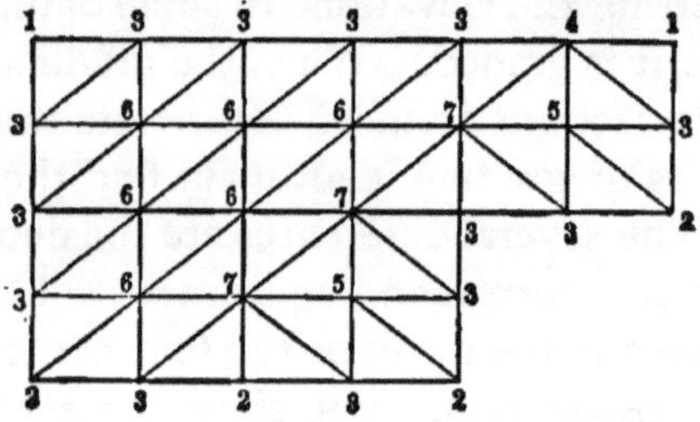

FIG. 109.

on a surface either diagonal may be drawn, which with the bounding lines makes two surfaces. If the ground is quite irregular, or if the rectangles are taken pretty large, the surveyor may note on the ground which diagonal would most nearly fit the surface. Let these be sketched in as shown in Fig. 109. Each rectangular area then becomes two triangles, and when computed as triangular prisms, each corner height at the end of a diagonal is used twice, while the two other corner heights are used but once. That is, twice as much weight is given to the corner heights on the diagonals as to the others. In Fig. 109, the same area as that in Fig. 108 is shown with the diagonals drawn which best fit the surface of the ground. The numbers at the corners indicate how many times each height is to be used. It will be seen that each height is used as many times as there are triangles meeting at that corner. To derive the formula for this case, take a single rectangle, as in Fig. 110, with the diagonal joining corners 2 and 4. Let A be the area of the rectangle. Then from the corollary, p. 395, we have for the volume of the rectangular prism, in cubic yards,

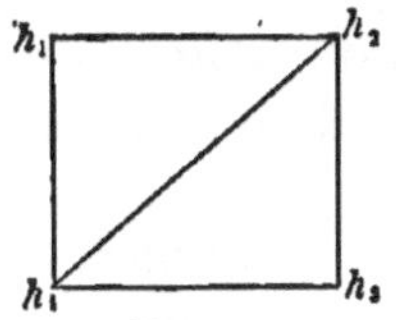

FIG. 110.

$$V = \frac{A}{2 \times 27}\left(\frac{h_1 + h_2 + h_4}{3} + \frac{h_2 + h_3 + h_4}{3}\right)$$

$$= \frac{A}{6 \times 27}(h_1 + 2h_2 + h_3 + 2h_4). \quad . \quad . \quad . \quad . \quad . \qquad (2)$$

For an assemblage of such rectangular prisms as shown in Fig. 109, the diagonals being drawn, we have, in cubic yards,

$$V = \frac{A}{6 \times 27}[\Sigma h_1 + 2\Sigma h_2 + 3\Sigma h_3 + 4\Sigma h_4 + 5\Sigma h_5$$

$$+ 6\Sigma h_6 + 7\Sigma h_7 + 8\Sigma h_8]; \quad . \quad . \quad . \qquad (3)$$

where A is the area of one rectangle, and the subscripts denote the number of triangles meeting at a corner.

As a check on the numbering of the corners, Fig. 109, add them all together and divide by six. The result should be the number of rectangles in the figure. In this case, if the rectangles be taken 36 feet by 45 feet, or, better, 40 feet by 40.5 feet, then the sum of the multiplied heights with the decimal point omitted is the number of cubic yards of earthwork, the corner heights having been taken out to tenths of a foot.

The method by diagonals is more accurate than that by rectangles simply, the dimensions being the same; or, for equal degrees of exactness larger rectangles may be used with diagonals than without them, and hence the work materially reduced. In any case some degree of approximation is necessary.

312. Approximate Estimates by means of Contours.—(*A*) Whenever an extended surface of irregular outline is to be graded down, or filled up to a given *plane* (not a warped or curved surface), a near approximation to the amount of cut or fill may be made from the contour lines. In Fig. 111 the full curved lines are contours, showing the original surface of the ground. Every fifth one is numbered, and these were the contours shown on the original plat. Intermediate contours one foot apart have been interpolated for the purpose of making this estimate. The figures around the outside of the bounding lines give the elevations of those points after it is graded down. The straight lines join points of equal elevation after grading; and since this surface is to be a plane these lines are surface or contour lines after grading. Wherever these two sets of contour lines intersect, the difference of their elevations is the depth of cut or fill at that point. If now we join the points of equal cut or fill (in this case it is all in cut), we obtain a new set of curves, shown in the figure by dotted lines, which may be used for estimating the amount of earthwork. The dotted boundaries are the horizontal projections of the traces on the natural surface of planes parallel to the final graded surface which are uniformly spaced one foot apart vertically. These projected areas are measured by the planimeter and called A_1, A_2, A_3, etc. Each area is bounded by the dotted line and the bounding lines of the figure, since on these

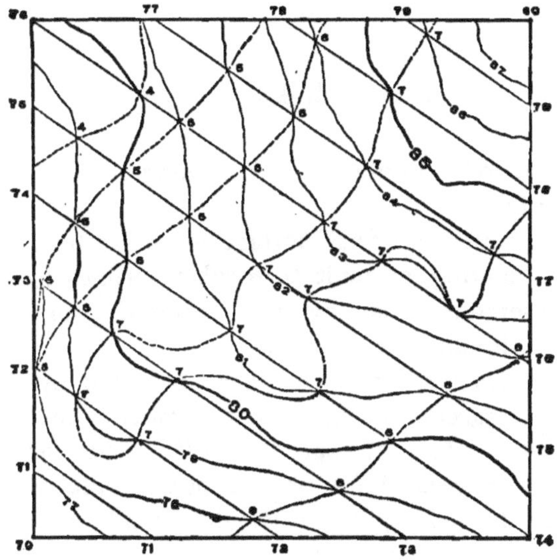

Fig. 111.

bounding lines all the projections of all the traces unite, the slope here being vertical. For any two adjoining layers we have, by the prismoidal formula* as well as by Simpson's one-third rule,

$$V_{1-3} = \frac{h}{3}(A_1 + 4A_2 + A_3), \quad . \quad . \quad . \quad . \quad . \quad (1)$$

where h is the common vertical distance between the projected areas.

* For the demonstration of the prismoidal formula see Art. 314.

For the next two layers we would have, similarly,

$$V_{3-5} = \frac{h}{3}(A_3 + 4A_4 A_5); \quad . \quad . \quad . \quad . \quad . \quad (2)$$

or for any even number of layers we would have, in cubic yards,

$$V = \frac{h}{3 \times 27}(A_1 + 4A_2 + 2A_3 + 4A_4 + 2A_5 + \quad . \quad . \quad . \quad . \quad A_n), \quad (3)$$

where n is an odd number, h and A being in feet and square feet respectively.

(B) Whenever the final surface is not to be a plane, but warped, undulating, or built to regular outlines like a fortification, a reservoir embankment, or terraced grounds, a different method should be employed.

In the former method the areas bounded by the dotted lines were areas cut out by planes parallel to the final plane surface, passed one foot apart *vertically*. But since the map shows only the *horizontal projections* of these planes, these projections, multiplied by the vertical distance between them, would give the true volumes.

When the final surface is not to be a plane, proceed as follows: First make a careful contour map of the ground. Then lay down on this map a system of contour lines, corresponding in elevation to the first set of contours, but in a different colored ink, which will accurately represent the final surface desired. This second set of contours would be a series of straight lines if a regular surface, composed of plane faces, was to be constructed, but would be curving lines if the ground were to be brought to a final curving or undulating surface.

The closed figures bounded by the two sets of intersecting contours of the same elevation are *horizontal* areas of cut or fill, separated by the common vertical distance between contours. The volumes here defined are oblique solids bounded by horizontal planes at top and bottom, and are a species of prismoid. The volume of one of these prismoids is found by applying the prismoidal formula to it, finding the end areas by means of a planimeter, and taking the length as the

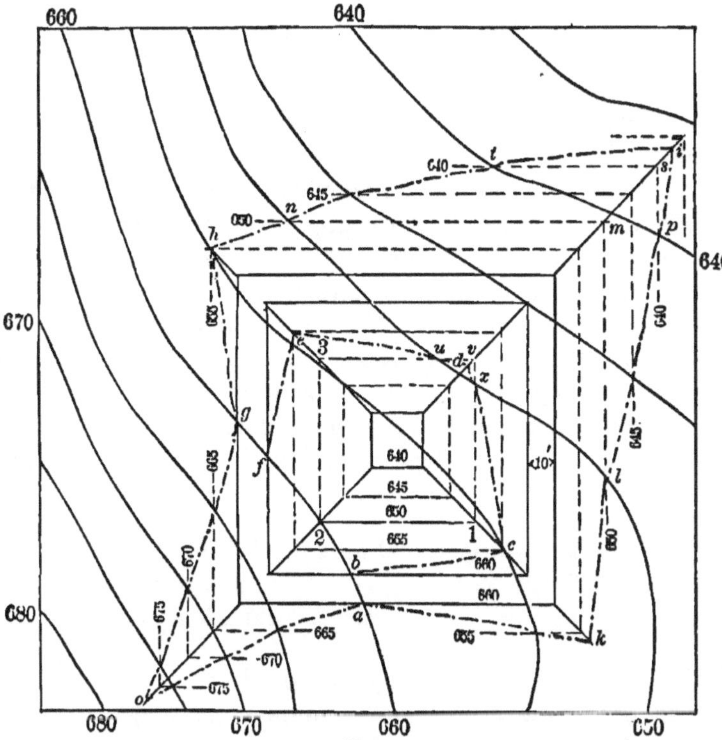

Fig. IIIa.

vertical distance between contours. If the contours be drawn close enough together, then each alternate contour-area may be used as a middle area, and the length of the prismoid taken at twice the vertical distance between contours; or the volume may be computed by either of the formulas (12), (13), (14), or (15) of Appendix C, where the *h's* would here become the end areas and *l* the vertical distance between contours.

Example: Let it be required to build a square reservoir on a hillside, which shall be partly in excavation and partly in embankment, the ground being such as shown by the full contour lines in Fig. 111*a*.*

The contours, for the sake of simplicity and brevity, are spaced five feet apart. The top of the wall, shown by the full lines making the square, is 10 feet wide and at an elevation of 660 feet. The reservoir is 20 feet deep, with side slopes, both inside and outside, of two to one, making the bottom elevation 640 feet, and 20 feet square, the top being 100 feet square on the inside. The dotted lines are contours of the finished slopes, both inside and out, at elevations shown on the figure. The areas in fill all fall within the broken line marked *a b c d e f g h i k*, and the cut areas all fall within the broken line marked *a b c d e f g o*. These broken lines are grade lines. The horizontal sectional areas in fill and cut are readily traced by following the closed figures formed by contours of equal elevation, thus—

At 640 foot level sectional area in fill is *p s t*.
" 650 " " " " " *l m n u v x l*.
" 650 " " " " cut is 1 2 3 *u x*.

The other areas are as easily traced. In the figure the lines have all been drawn in black. In practice they should be drawn in different colors to avoid confusion.

This second method should be used in all cases where the graded area is considerable and the final relief form is not a plane. If the contours be carefully determined and be taken

* This figure is taken from a paper describing the method by Prof. William G. Raymond, University of California.

near enough together, the method will give as accurate results as may be obtained in any other way. The volume may be computed by eq. (3) of this article, where the areas are the horizontal sectional areas bounded by contours of equal elevation, and h is the vertical distance between contours.

When these methods are used for final estimates, the contours should be carefully determined, and spaced not more than two feet apart on steep slopes and one foot apart on low slopes.

313. **The Prismoid** is a solid having parallel end areas, and may be composed of any combination of prisms, cylinders, wedges, pyramids, or cones or frustums of the same, whose bases and apices lie in the end areas. It may otherwise be defined as a volume generated by a right-line generatrix moving on the bounding lines of two closed figures of any shapes which lie in parallel planes as directrices, the generatrix not necessarily moving parallel to a plane director. Such a solid would usually be bounded by a warped surface, but it can always be subdivided into one or more of the simple solids named above.

Inasmuch as cylinders and cones are but special forms of prisms and pyramids, and warped surface solids may be divided into elementary forms of them, and since frustums may also be subdivided into the elementary forms, it is sufficient to say that all prismoids may be decomposed into prisms, wedges, and pyramids. If a formula can be found which is equally applicable to all of these forms, then it will apply to any combination of them. Such a formula is called

314. **The Prismoidal Formula.**

Let A = area of the base of a prism, wedge, or pyramid;

A_1, A_m, A_2 = the end and middle areas of a prismoid, or of any of its elementary solids;

h = altitude of the prismoid or elementary solid.

Then we have,
For Prisms,

$$V = hA = \frac{h}{6}(A_1 + 4A_m + A_2). \quad . \quad . \quad . \quad . \quad (1)$$

For Wedges,

$$V = \frac{hA}{2} = \frac{h}{6}(A_1 + 4A_m + A_2). \quad . \quad . \quad . \quad . \quad (2)$$

For Pyramids,

$$V = \frac{hA}{3} = \frac{h}{6}(A_1 + 4A_m + A_2). \quad . \quad . \quad . \quad . \quad (3)$$

Whence for any combination of these, having all the common altitude *h*, we have

$$V = \frac{h}{6}(A_1 + 4A_m + A_2), \quad . \quad . \quad . \quad . \quad . \quad . \quad . \quad (4)$$

which is the prismoidal formula.

It will be noted that this is a rigid formula for all prismoids. The only approximation involved in its use is in the assumption that the given solid may be generated by a right line moving over the boundaries of the end areas.

This formula is used for computing earthwork in cuts and fills for railroads, streets, highways, canals, ditches, trenches, levees, etc. In all such cases, the shape of the figure above the natural surface in the case of a fill, or below the natural surface in the case of a cut, is previously fixed upon, and to complete the closed figure of the several cross-section areas only the outline of the natural surface of the ground at the section remains to be found. These sections should be located so near together that the intervening solid may fairly be assumed to be a prismoid. They are usually spaced 100 feet apart, and then intermediate sections taken if the irregularities seem to require it.

The *area* of the middle section is never the mean of the two end areas if the prismoid contains any pyramids or cones among its elementary forms. When the three sections are similar in form, the *dimensions* of the middle area are always the means of the corresponding end dimensions. This fact often enables the dimensions, and hence the area of the middle section, to be computed from the end areas. Where this cannot be done, the middle section must be measured on the ground, or else each alternate section, where they are equally spaced, is taken as a middle section, and the length of the prismoid taken as twice the distance between cross-sections. For a continuous line of earthwork, we would then have, in cubic yards,

$$V=\frac{l}{3\times 27}(A_1+4A_2+2A_3+4A_4+2A_5+4A_6 \ldots +A_n), \quad (1)$$

where l is the distance between sections in feet. This is the same as equation (3), p. 401. Here the assumption is made that the volume lying between alternate sections conforms sufficiently near to the prismoidal forms.

315. Areas of Cross-sections.—In most cases, in practice at least, three sides of a cross-section are fixed by the conditions of the problem. These are the side slopes in both cuts and fills, the bottom in cuts and the top in embankments, or fills. It then remains simply to find where the side slopes will cut the natural surface, and also the form of the surface line on the given section. Inasmuch as stakes are usually set at the points where the side slopes cut the surface, whether in cut or fill, such stakes are called slope-stakes, and they are set at the time the cross-section is taken. The side slopes are defined as so much horizontal to one vertical. Thus a slope of $1\frac{1}{2}$ to 1 means that the horizontal component of a given portion of a slope-line is $1\frac{1}{2}$ times its vertical component, the horizontal component always being named first. The *slope-ratio* is the ratio of the horizontal to the vertical component, and is therefore always the same as the first number in the slope-definition. Thus for a slope of $1\frac{1}{2}$ to 1 the slope-ratio is $1\frac{1}{2}$.

316. The Centre and Side Heights.—The centre heights are found from the profile of the surface along the centre line, on which has been drawn the grade line of the proposed work. These are carefully drawn on cross-section paper, when the height of grade at each station above or below the surface line can be taken off. These centre heights, together with the width of base and side slopes in cuts and in fills, are the necessary data for fixing the position of the slope-stakes. When these are set for any section as many points on the surface line joining them may be taken as desired. In ordinary rolling ground usually no intermediate points are taken, the centre point being already determined. In this case three points in the surface line are known, both as to their distance out from the centre line and as to their height above the grade line. Such sections are called "three-level sections," the surface lines being assumed straight from the slope-stakes to the centre stake.

317. The Area of a Three-level Section.

Let d and d' be the distances out, and
h and h' the heights above grade of right and left slope-stakes, respectively;
D the sum of d and d',
c the centre height,
r the slope-ratio,
w the width of bed.

Then the area $ABCDE$ is equal to the sum of the four triangles AEw, BCw, wCD, and wED. Or,

$$A = \frac{(d + d')\,c + (h + h')\frac{w}{2}}{2}. \quad . \quad . \quad . \quad . \quad . \quad (1)$$

This area is also equal to the sum of the triangles FCD and FED, minus the triangle AFB. Or,

$$A = \left(c + \frac{w}{2r}\right)\frac{D}{2} - \frac{w^2}{4r}. \quad . \quad . \quad . \quad . \quad . \quad (2)$$

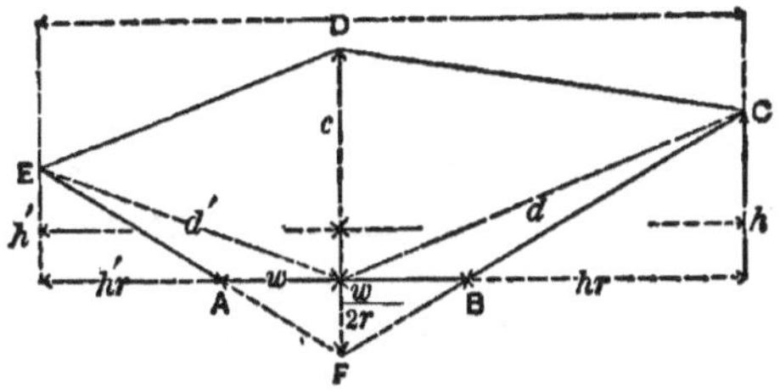

FIG. 112.

Equation (2) can also be obtained directly from equation (1) by substituting for h and h' in (1) their values in terms of d and w, $h = \frac{d - \frac{w}{2}}{r}$, and then putting $D = d + d'$. Equation (2) has but two variables, c and D, and is the most convenient one to use.

318. Cross-sectioning.—It will be seen from Fig. 112 that in the case of a three-level section the only quantities to be determined in the field are the heights, h and h', and the distances out, d and d', of the slope-stakes. These are found by trial. A levelling instrument is set up so as to read on the three points C, D, E, and the rod held first at D. The reading here gives the height of instrument above this point. Add this algebraically to the centre height (which may be negative, and which has been obtained from the profile for each station), and the sum is the height of instrument above (or below) the grade line. If the ground were level transversely, the distance out to the slope-stakes would be

$$d = cr + \frac{w}{2}.$$

But this is not usually the case, and hence the distance out must be found by trial. If the ground slopes $\left\{ \begin{matrix} \text{down} \\ \text{up} \end{matrix} \right\}$ from the centre line in a $\left\{ \begin{matrix} \text{fill} \\ \text{cut} \end{matrix} \right\}$ the distance out will evidently be more than that given by the above equation, and *vice versa*. The rodman estimates this distance, and holds his rod at a certain measured distance out, d_1. The observer reads the rod, and deducts the reading from the height of instrument above grade (or adds it to the depth of instrument below grade), and this gives the height of that point, h_1, above or below grade. Its distance out, then, *should* be $d = h_1 r + \frac{w}{2}$. If this be more than the actual distance out, d_1, the rod is set farther out; if less, it is moved in. The whole operation is a very simple one in practice, and the rodman soon becomes very expert in estimating nearly the proper position the first time.

In heavy work—that is, for large cuts or fills, and for irregular ground—it may be necessary to take the elevation and distance out of other points on the section in order to better determine its area. These are taken by simply reading on the rod at the critical points in the outline, and measuring the distances out from the centre. The points can then be plotted on cross-section paper and joined by straight or by free-hand curved lines. In the latter case the area should be determined by planimeter.

319. Three-level Sections, the Upper Surface consisting of two Warped Surfaces.—If the three longitudinal lines joining the centre and side heights on two adjacent three-level sections be used as directrices, and two generatrices, one on each side the centre, be moved parallel to the end areas as plane directers, two warped surfaces are generated, every cross-section of which parallel to the end areas is a three-level section. These same surfaces could be generated by two longitudinal generatrices, moving over the surface end-area lines as directrices. The surface would therefore be a prismoid, and its exact volume would be given by the prismoidal formula. *The middle area* in this case is readily found, since the center and side heights are the means of the corresponding end dimensions.

The prismoidal formula, giving volumes in cubic yards,

$$V = \frac{l}{6 \times 27}(A_1 + 4A_m + A_2), \quad . \quad . \quad . \quad . \quad (1)$$

could therefore be written

$$V = \frac{l}{12 \times 27}\left[\left(c_1 + \frac{w}{2r}\right) D_1 + \left(c_2 + \frac{w}{2r}\right) D_2 + 4\left(c_m + \frac{w}{2r}\right) D_m\right] - \frac{lw^2}{4 \times 27r}. \quad . \quad . \quad (2)$$

This equation is derived directly from eq. (1) above, and eq. (2), p. 406. The quantity $\frac{w}{2r}$ is the distance from the grade-plane to the intersection of the side slopes, and is a constant for any given piece of road. It would have different values, however, in cuts and fills on the same line.

For brevity, let

$$\frac{w}{2r} = c_0; \quad \text{and} \quad \frac{lw^2}{4 \times 27r} = \frac{lwc_0}{54} = K.$$

Here K is the volume of the prism of earth, 100 feet long, included between the roadbed and side slopes. It is first included in the computation and then deducted. It is also a constant for a given piece of road.

Equation (2) now becomes

$$V = \frac{l}{12 \times 27}[(c_1 + c_0)D_1 + (c_2 + c_0)D_2 + 4(c_m + c_0)D_m] - K, \quad (3)$$

where c_m and D_m are the means of $c_1 c_2$ and $D_1 D_2$, respectively.

This equation involves but two kinds of variables, c and D, and is well adapted to arithmetical, tabular, or graphical computation. Thus if $l = 100$; $w = 18$; and $r = 1\frac{1}{2}$; then $c_0 = 6$; and $K = 200$; and equation (3) becomes

$$V = \tfrac{100}{324}[(c_1 + 6)D_1 + (c_2 + 6)D_2 + 4(c_m + 6)D_m] - 200 \quad (4)$$

If the total centre heights (to intersection of side slopes) be represented by C_1, C_2, and C_m, then eq. (3) becomes, in general,

$$V = K'(C_1D_1 + C_2D_2 + 4C_mD_m) - K, \quad . \quad . \quad . \quad (5)$$

where $K' = \frac{100}{324}$, and is independent of width of bed and of slopes.

For any given piece of road, the constants K, K', and c_0 are known, and for each prismoid the C's and D's are observed, hence for any prismoid all the quantities in eq. (5) are known.

320. Construction of Tables for Prismoidal Computation.—If a table were prepared giving the products $K'CD$ for various values of C and D, it could be used for evaluating equation (3), which is the same as equation (5). The arguments would be the total widths (D_1), and the centre heights (C_1). Such a table would have to be entered three times for each prismoid, first with C_1 and D_1; second with C_2 and D_2; and finally with C_m and D_m. If four times the last tabular value be added to the sum of the other two, and K subtracted, the result is the true volume of the prismoid.

VALUES OF $c_0 \left(= \frac{w}{2r}\right)$ AND $K \left(= \frac{lw^2}{4 \times 27r}\right)$ FOR VARIOUS WIDTHS AND SLOPES.

Width of Road-bed.	Slopes.															
	¼ to 1.		½ to 1.		¾ to 1.		1 to 1.		1¼ to 1.		1½ to 1.		1¾ to 1.		2 to 1.	
	C_0	K	C_0	K	C_0	K	C_0	K	C_0	K	C_0	K	C_0	K	C_0	K
10	20	370	10	185	6.7	123	5.0	93	4.0	74	3.3	62	2.9	53	2.5	46
11	22	448	11	224	7.3	149	5.5	112	4.4	90	3.7	75	3.1	64	2.8	56
12	24	533	12	266	8.0	178	6.0	133	4.8	107	4.0	89	3.4	76	3.0	67
13	26	626	13	313	8.7	209	6.5	157	5.2	125	4.3	104	3.7	89	3.2	78
14	28	725	14	363	9.3	242	7.0	181	5.6	145	4.7	121	4.0	104	3.5	91
15	30	833	15	417	10.0	278	7.5	208	6.0	167	5.0	139	4.3	119	3.8	104
16	32	948	16	471	10.7	316	8.0	237	6.4	190	5.3	158	4.6	135	4.0	118
17	34	1070	17	535	11.3	357	8.5	268	6.8	214	5.7	178	4.9	153	4.2	134
18	36	1200	18	600	12.0	400	9.0	300	7.2	240	6.0	200	5.1	171	4.5	150
19	38	1337	19	668	12.7	446	9.5	334	7.6	267	6.3	223	4.4	191	4.8	167
20	40	1481	20	740	13.3	494	10.0	370	8.0	296	6.7	247	5.7	212	5.0	185
21	42	1633	21	816	14.0	544	10.5	408	8.4	327	7.0	272	6.0	233	5.2	204
22	44	1793	22	896	14.7	598	11.0	448	8.8	359	7.3	299	6.3	256	5.5	224
23	46	1959	23	980	15.3	653	11.5	490	9.2	392	7.7	326	6.6	280	5.8	245
24	48	2134	24	1067	16.0	711	12.0	534	9.6	427	8.0	356	6.9	305	6.0	267
25	50	2315	25	1158	16.7	772	12.5	579	10.0	463	8.3	386	7.1	331	6.2	264
26	52	2504	26	1252	17.3	835	13.0	626	10.4	501	8.7	417	7.4	358	6.5	313
27	54	2700	27	1350	18.0	900	13.5	675	10.8	540	9.0	450	7.7	386	6.8	338
28	56	2904	28	1452	18.7	968	14.0	726	11.2	581	9 3	484	8.0	415	7.0	363
29	58	3115	29	1558	19.3	1038	14.5	779	11.6	623	9.7	519	8 3	445	7.2	389
30	60	3333	30	1667	20.0	1111	15.0	833	12.0	667	10.0	556	8.6	476	7.5	417

Table XI.* is such a table, computed for total centre heights from 1 to 50 feet, and for total widths from 1 to 100 feet. In railroad work neither of these quantities can be as small as one foot, but the table is designed for use in all cases where the parallel end areas may be subdivided into an equal number of triangles or quadrilaterals.

EXAMPLE 1. *Three-level Ground having two Warped Surfaces.*—Find the volume of two prismoids of which the following are the field-notes, the width of bed being 20 feet, and the slopes $1\frac{1}{2}$ to 1.

Station 11.	$\frac{28.9\dagger}{+12.6}$	$\frac{0}{+18.6}$	$\frac{43.0}{+22.0}$
Station 12.	$\frac{27.1}{+11.4}$	$\frac{0}{+14.8}$	$\frac{40.3}{+20.2}$
Station 12 + 56.	$\frac{24.3}{+9.5}$	$\frac{0}{+10.3}$	$\frac{34.9}{+16.6}$

From the table, p. 410, giving values of C_0 and K, we find for $w = 20$, and $r = 1\frac{1}{2}$, $C_0 = 6.7$, and $K = 247$.

The computation may be tabulated as follows:

Sta.	Width, $D=d+d'$.	Height, $C=c+c_0$.	Partial Volume.	Volume of Prismoid.
11	71.9	25.3	562	
M	69.6	23.4	503 × 4 = 2012	
12	67.4	21.5	447	
			3021 − 247	2774
M	63.3	19.2	374 × 4 = 1496	
12 + 56	59.2	17.0	311	
			.56 (2254 − 247)	1124

* Modeled somewhat after Crandall's Tables, but adapted to give volumes by the Prismoidal Formula at once instead of by the method of mean end areas first and correcting by the aid of another table to give prismoidal volumes, as Prof. Crandall has done.

† The numerators are the distances out, and the denominators are the heights above grade, + denoting cut and − fill.

Entering the table (No. XI.) for a width of 71 and a height of 25, we find 548, to which add 7 for the 3 tenths of height, and 7 more for the 9 tenths in width, both mentally, thus giving 562 cu. yds. for this partial volume. Similarly for the width 67.4, and height 21.5, obtaining 447 cu. yds. The corresponding result for the middle area is 503, which is to be multiplied by 4, thus giving 2012 cu. yds. The sum of these is 3021 cu. yds., from which is to be subtracted the constant volume K, which in this case is 247 cu. yds., leaving 2774 cu. yds. as the volume of the prismoid.

The next prismoid is but 56 feet long, but it is taken out just the same as though it were full, and then 56 hundredths of the resulting volume taken. The data for the 12th station is used in getting this result without writing it again on the page.

EXAMPLE 2. *Five-level Ground having four Warped Surfaces.*—Find the volume of a prismoid of which the following are the field-notes, the width of bed being 20 feet, and the slopes $1\frac{1}{2}$ to 1:

11.	$\frac{28.9}{+12.6}$	$\frac{15.0}{+12.0}$	$\frac{0}{+18.6}$	$\frac{20.0}{+21.0}$	$\frac{43.0}{+22.0}$
12.	$\frac{27.1}{+11.4}$	$\frac{12.5}{+12.0}$	$\frac{0}{+14.8}$	$\frac{18.5}{+19.6}$	$\frac{40.3}{+20.2}$

This is the same problem as the preceding, with intermediate heights added.

To compute this from the table, it is separated into three prismoids, as shown in Fig. 113.

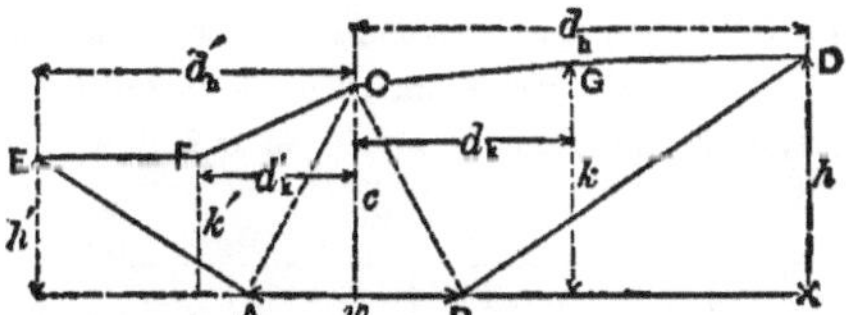

FIG. 113.

Let $ABDGCFE$ be the cross-section. This may be separated into the triangle ABC, and the two quadrilaterals $BCGD$ and $ACFE$. The area of the triangle is $\frac{1}{2}cw$. That of the right quadrilateral is, from Art. 179, p. 202,

$$\tfrac{1}{2}\left[c\left(d_k - \frac{w}{2}\right) + k(d_h - 0) + h\left(\frac{w}{2} - d_k\right)\right] = \tfrac{1}{2}\left[(c - h)\left(d_k - \frac{w}{2}\right) + kd_h\right].$$

Similarly the area of the left quadrilateral is $\tfrac{1}{2}\left[(c - h')\left(d'_k - \frac{w}{2}\right) + k'd'_h\right]$.

The total area of the section then is

$$A = \tfrac{1}{2}\left[(c - h')\left(d'_k - \frac{w}{2}\right) + k'd'_h + cw + kd_h + (c - h)\left(d_k - \frac{w}{2}\right)\right]. \quad . \quad . \quad (1)$$

If the interior side elevations be taken over the edges of the base, then $d'_k - \frac{w}{2}$ and $d_k - \frac{w}{2}$ both become zero, and the first and last terms disappear. Or if the centre and extreme side heights are the same, these terms go out. Experience shows that these terms can usually be neglected without material error. If they are retained, each partial volume will be composed of five terms, while if they are neglected there will be but three. The signs of these terms also must be carefully attended to. When the interior side readings are taken over the edges of the base, therefore, this equation becomes

$$A = \tfrac{1}{2}(k'd'_h + cw + kd_h) \quad . \quad . \quad . \quad . \quad . \quad . \quad . \quad . \quad . \quad (2)$$

The tables are well adapted to compute the prismoidal volume for five-level sections by either of these formulæ. Thus, if the adjacent section also has five points determined in its surface, its area may be represented by an equation similar to one of these, and from these end-area data mean values may be found for the corresponding middle-area points, and the volumes taken out as before. In this case the prism included between the road-bed and side-slopes, whose volume is K, is not included, and hence its volume is not to be deducted from the result. The computation by table XI. of equation (1) would be as follows:

Sta.	h'.	d'_h.	k'.	d'_k.	c.	d_k.	k.	d_h.	h.	Partial Volumes.	Total Volume.
11	12.6	28.9	12.0	15.0	18.6	20.0	21.0	43.0	22.0	+9+108+114+279−10 = 500	
M	12.0	28.0	12.0	13.8	16.7	19.2	20.3	41.6	21.1	4(+6+104+102+260−12)=1840	
12	11.4	27.1	12.0	12.5	14.8	18.5	19.6	40.3	20.2	+3+100+ 90+242−13 = 422	2762

The use of the table is the same as before. First take out from the table the volume corresponding to $(c - h')\left(d'_h - \frac{w}{2}\right)$, which when evaluated for section 11 is $(18.6 - 12.6)(15.0 - 10) = 6.0 \times 5.0$. This is positive, and the volume corresponding to a depth of 6.0 feet and a width of 5.0 feet is 9 cubic yards. Proceed to evaluate the remaining terms of eq. (1) in a similar manner, the last term coming out negative. The dimensions of the mid section are the means of the corresponding end dimensions, as before. If one end-area is a three-level section and the next a five-level section, the included prismoid is computed as a five-level prismoid, the vanishing points in the three-level section corresponding to the interior side elevations on the five-level section being indicated in the field. Partial stations, or prismoids, are first computed as though they were 100 feet long (for which the table is constructed), and then multiplied by their length and divided by 100 as before.

If equation (2) may be used, the work is shortened very much. The columns in h', d'_h, d_h, and h, may be omitted, and there will also be but three terms in each partial product. Thus, if sections 11 and 12 had been taken with the interior elevations, each 10 feet from the centre line, we might have had something as follows:

$$11. \quad \frac{28.9}{+12.6} \quad \frac{10.0}{+15.4} \quad \frac{0}{+18.6} \quad \frac{10.0}{+19.8} \quad \frac{43.0}{+22.0}$$

$$12. \quad \frac{27.1}{+11.4} \quad \frac{10.0}{+12.5} \quad \frac{0}{+14.8} \quad \frac{10.0}{+17.4} \quad \frac{40.3}{+20.2}$$

The computation then, by eq. (2), would have been:

Sta.	d'_h.	h'.	c.	h.	d_h.	Partial Volumes.	Total Volume.
11	28.9	15.4	18.6	19.8	43.0	137 + 114 + 263 = 514	
M	28.0	14.0	16.7	18.6	41.6	4 (121 + 102 + 239) = 1848	
12	27.1	12.5	14.8	17.4	40.3	104 + 90 + 215 = 409	2771

By this method the computation of a five-level section is little more trouble than that of a three-level section, and yet the intermediate points taken at a distance of $\frac{w}{2}$ from the centre, are apt to increase the accuracy considerably on ordinary rolling ground.

321. Three-level Sections, the Surface divided into four Planes by Diagonals.—If the surface included between two three-level sections be assumed to be made up of four planes formed by joining the centre height at one end with a side height at the other end section on each side the centre line (Fig. 114), these lines being called diagonals, an exact computation of the volume is readily made without computing the mid-area. Two diagonals are possible on each side the centre line but the one is drawn which is observed to most nearly fit the surface. They are noted in the field when the cross-sections are taken.

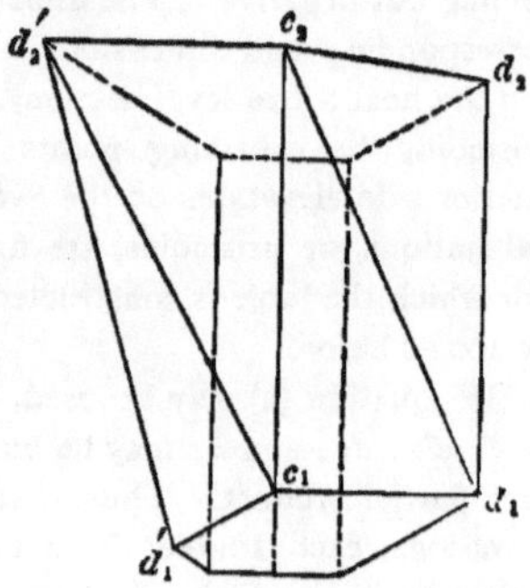

FIG. 114.

The total volume of such a prismoid in cubic * yards is

$$V = \frac{l}{6 \times 27}\Big[(d_1 + d_1')c_1 + (d_2 + d_2')c_2 + DC + D'C' + \frac{w}{2}(h_1 + h_2 + H + h_1' + h_2' + H')\Big],^* \quad (1)$$

where c_1, h_1, and h_1' are the centre and side heights at one section and d_1 and d_1' the distances out, c_2, h_2', h_2, d_2, and d_2' be-

* For a demonstration of this formula see Henck's Field-Book.

ing the corresponding values for the other end section. *C* and *C'* are the centre heights, *H* and *H'* the side heights, and *D* and *D'* the distances out on the right and left diagonals. Although this formula seems long, the computations by it are very simple. Thus let the volume be found from the following field-notes for a base of 20 feet and side slopes $1\frac{1}{2}$ to 1.

A_1	$\frac{22}{+8}$	$\frac{0}{+8}$	$\frac{47.5}{+25}$.
A_2	$\frac{34}{+16}$	$\frac{0}{+4}$	$\frac{16}{+4}$.

The upper figures indicate the distances out and those below the lines the heights, the plus sign being used for cuts. The computation in tabular form is as follows:

Sta.	*d.*	*h.*	*c.*	*h'.*	*d'.*	*d* + *d'*.	(*d* + *d'*)*c.*	*DC.*	*D'C'.*
1	22	8	8	25	47.5	69.5	556	....	....
2	34	16	4	4	16	50.0	200	88	128
		$h_1 + h_2 = 24$					88		
		$H + H' = 12$					128		
		$\frac{w}{2}\Sigma h$'s = 65 × .o					= 650		
							6) 162200		
							27) 27033		
							1001 cu. yards.		

The great advantage of the method consists in the data all being at hand in the field-notes.

Hudson's Tables * give volumes for this kind of prismoid.

* Tables for Computing the Cubic Contents of Excavations and Embankments. By John R. Hudson, C.E. John Wiley & Sons, New York, 1884.

They furnish a very ready method of computing volumes when this system is used.

322. **Comparison of Methods by Diagonals and by Warped Surfaces.**—Although the surveyor has a choice of two sets of diagonals when this method is used, the real surface would usually correspond much nearer the mean of the two pairs of plane surfaces than to either one of them. That is, the natural surface is curved and not angular, and therefore it is probable that two warped surfaces joining two three-level sections would generally fit the ground better than four planes, notwithstanding the choice that is allowed in the fitting of the planes. More especially must this be granted when the truth of the following proposition is established.

PROPOSITION: *The volume included between two three-level sections having their corresponding surface lines joined by warped surfaces, is exactly a mean between the two volumes formed between the same end sections by the two sets of planes resulting from the two sets of diagonals which may be drawn.*

If the two sets of diagonals be drawn on each side the centre line and a cross-section be taken parallel to the end areas, the traces of the four surface planes on each side the centre line on the cutting plane will form a parallelogram, the diagonal of which is the trace of the warped surface on this cutting plane. Since this cutting plane is any plane parallel to the end areas, and since the warped surface line bisects the figure formed by the two sets of planes formed by the diagonals, it follows that the warped surface bisects the volume formed by the two sets of planes. The proposition will therefore be established if it be shown that the trace of the warped surface is the diagonal of the parallelogram formed by the traces of the four planes formed by the two sets of diagonals. Fig. 115 shows an extreme case where the centre height is higher than the side height at one end and lower at the other. Only the left half of the prismoid is shown in the figure. The cutting plane cuts the centre and side lines and the two diagonals in *efgh* on the plane, and in $e'f'g'h'$ on the vertical projection. For the diagonal c_1d_2 the surface lines cut out are $e'f'$ and $f'h'$. For the diagonal c_2d_1 they are $e'g'$ and $g'h'$. For the warped surface the line cut out is $e'h'$, this being an

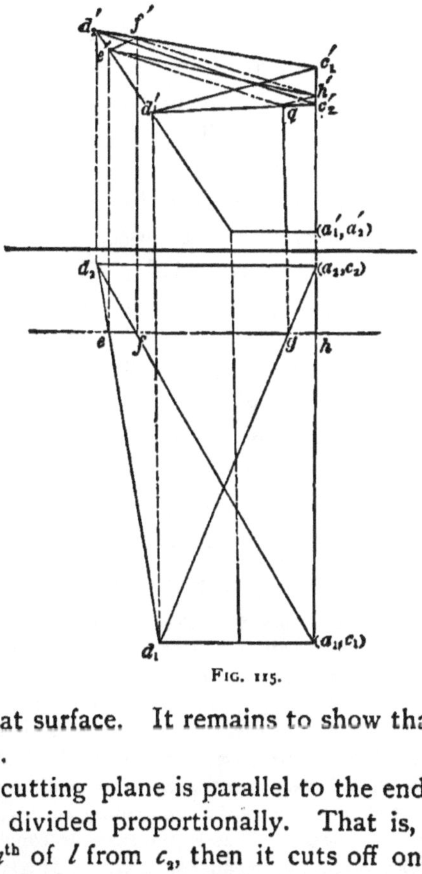

FIG. 115.

element of that surface. It remains to show that $e'f'h'g'$ is a parallelogram.

Since the cutting plane is parallel to the end planes all the lines cut are divided proportionally. That is, if the cutting plane is one n^{th} of l from c_2, then it cuts off one n^{th} of all the lines cut, measured from that end plane. But if the lines are divided proportionally, the projections of those lines are divided proportionally, and hence the points e', f', h', g' divide the sides of the quadrilateral d_2', c_1', c_2', d_1' proportionally. But it is a proposition in geometry that if the four sides of a quadrilateral, or two opposite sides and the diagonals, be divided proportionally and the corresponding points of subdivision joined, the resulting figure is a parallelogram. Therefore $e'f'h'g'$ is a parallelogram, and $e'h'$ is one of its diagonals and hence bisects it. Whence the surface generated by this line moving along c_1c_2 and d_1d_2 parallel to the end areas bisects the volume formed by the four planes resulting from the use of both diagonals on one side the centre line. Q. E. D.

It is probable, therefore, that the warped surface would usually fit the ground better than either of the sets of planes formed by the diagonals. Furthermore, the errors caused by the use of the warped surface (Table XI.) are compensating errors, thus preventing any marked accumulation of errors in a series of prismoids.* There are extreme cases, however, such as that given in the example, Fig. 114, which are best computed by the method by diagonals.

323. Preliminary Estimate from the Profile.—If the cross-sections be assumed level transversely then for given width of bed and side slopes, a table of end areas may be prepared in terms of the centre heights. From such a table the

* The two methods here discussed are the only ones that have any claims to accuracy. The method by "mean end areas," wherein the volume is assumed to be the mean of the end areas into the length, always gives too great a volume (except when a greater centre height is found in connection with a less total width, which seldom occurs), the excess being *one sixth of the volume of the pyramids involved in the elementary forms of the prismoid.* This is a large error even in level sections, and very much greater on sloping ground, and yet it is the basis of most of the tables used in computing earthwork, and in some States it is legalized by statute. Thus in the example computed by Henck's method on p. 414 the volume by mean end areas is 1193 cu. yards; by the prismoidal formula it is 1168 cu. yards, while by the method by diagonals it was only 1001 cu. yards. This was an extreme case, however, and was selected to show the adaptation of the method by diagonals to such a form.

end areas may be rapidly taken out and *plotted as ordinates from the grade line.* The ends of these ordinates may then be joined by a free-hand curve, and the area of this curve found by the planimeter. The ordinates may be plotted to such a scale that each unit of the area, as one square inch, shall represent a convenient number of cubic yards, as 1000. The record of the planimeter then in square inches and thousandths gives at once the cubic yards on the entire length of line worked over by simply omitting the decimal point. Evidently the scale to which the ordinates are to be drawn to give such a result is not only a function of the width of bed and side slopes, but also of the longitudinal scale to which the profile line is plotted. The area of a level section is

$$A = wc + rc^2, \quad . \quad . \quad . \quad . \quad . \quad . \quad . \quad (1)$$

where w, c, and r are the width of base, centre height, and slope-ratio respectively.

Now if h = the horizontal scale of the profile, that is the number of feet to the inch, and if one square inch of area is to represent 1000 cu. yards, the length of the ordinate must be

$$y = \frac{hA}{1000 \times 27} = \frac{h(wc + rc^2)}{27,000}. \quad . \quad . \quad . \quad . \quad . \quad (2)$$

If values be given to h, w, and r, which are constants for any given case, then the value of y becomes a function of c only, and a table can be easily prepared for the case in hand. Since y is a function of the second power of c, the second difference will be a constant, and the table can be prepared by means of first and second differences. Thus if c takes a small increment, as 1 foot, then the first difference is

$$\Delta' y = \frac{h}{27,000}(w + 2rc + r). \quad . \quad . \quad . \quad . \quad . \quad (3)$$

But this first difference is also a function of c, and hence when c takes an increment this first difference changes by an amount equal to

$$\Delta'' y = \frac{h}{27000} \cdot 2r, \quad . \quad . \quad . \quad . \quad . \quad . \quad (4)$$

which is constant. An initial first difference being given for a certain value of c, a column of first differences can be obtained by simply adding the $\Delta'' y$ continuously to the preceding sum. With this column of first differences the corresponding column of values of y may be found by adding the first differences continuously to the initial value of y for that column.*

TABULAR VALUES OF y IN EQUATION (2) FOR $w = 20$, $r = 1\frac{1}{2}$, AND $h = 400$.

c	0.′0	0.′1	0.′2	0.′3	0.′4	0.′5	0.′6	0.′7	0.′8	0.′9
	in.	in.	in.	in.	in.	in.	in.	in.	in.	in.
0	0.00	0.03	0.06	0.09	0.12	0.15	0.19	0.22	0.25	0.28
1	.32	.35	.39	.42	.46	.49	.53	.57	.61	.64
2	.68	.72	.76	.80	.84	.88	.92	.96	1.00	1.05
3	1.09	1.13	1.17	1.22	1.26	1.31	1.35	1.40	1.45	1.49
4	1.54	1.59	1.63	1.69	1.73	1.78	1.83	1.88	1.93	1.99
5	2.04	2.09	2.14	2.19	2.24	2.30	2.36	2.41	2.47	3.52
6	2.58	2.63	2.69	2.75	2.80	2.87	2.92	2.98	3.04	3.10
7	3.16	3.22	3.28	3.35	3.41	3.47	3.54	3.60	3.66	3.73
8	3.79	3.86	3.92	3.99	4.05	4.13	4.19	4.26	4.33	4.40
9	4.47	4.54	4.60	4.68	4.75	4.82	4.89	4.97	5.04	5.11
10	5.18	5.26	5.33	5.40	5.48	5.56	5.64	5.72	5.79	5.87
11	5.95	6.03	6.10	6.18	6.26	6.35	6.43	6.51	6.59	6.67
12	6.76	6.84	6.92	7.00	7.09	7.18	7.26	7.35	7.43	7.52
13	7.61	7.70	7.78	7.86	7.96	8.05	8.14	8.23	8.32	8.41
14	8.50	8.60	8.68	8.77	8.87	8.97	9.06	9.16	9.25	9.35
15	9.44	9.54	9.63	9.73	9.83	9.94	10.03	10.13	10.23	10.33
16	10.43	10.53	10.62	10.73	10.83	10.94	11.04	11.15	11.25	11.35
17	11.46	11.56	11.66	11.77	11.88	12.00	12.10	12.21	12.31	12.42
18	12.53	12.64	12.75	12.86	12.97	13.09	13.20	13.32	13.42	13.54
19	13.65	13.77	13.87	13.99	14.10	14.23	14.34	14.47	14.58	14.70
20	14.81	14.93	15.04	15.16	15.29	15.[illegible]	15.53	15.66	15.78	15.90

* For a further exposition of this subject, see Appendix C.

The preceding table was constructed in this manner, for $w = 20$ feet, $r = 1\frac{1}{2}$; and $h = 400$ feet to the inch.

324. Borrow-pits are excavations from which earth has been "borrowed" to make an embankment. It is generally preferable to measure the earth in cut rather than in fill, hence when the earth is taken from borrow-pits and its volume is to be computed in cut, the pits must be carefully staked out and elevations taken both before and after excavating. The methods given in art. 311 are well suited to this purpose, or they may be computed as prismoids by the aid of Table XI., if preferred. To use the table it is only necessary to enter it with such heights and widths as give twice the elementary areas (triangles or quadrilaterals) into which the end sections are divided, and then multiply the final result by the length and divide by 100. The table is entered for both end-area dimensions and also the mid-area dimensions, four times this latter result being taken the same as before.

325. Shrinkage of Earthwork.—Excavated earth first increases in volume, when removed from a cut and dumped on a fill, but it gradually settles, or shrinks, until it finally comes to occupy a less volume than it formerly did in the cut. Both the amounts, initial increase, and final shrinkage depend on the nature of the soil, its condition when removed, and the manner of depositing it in place. There can therefore be no general rules given which will always apply. *For ordinary clay and sandy loam, dumped loosely, the first increase is about one twelfth, and then the settlement about one sixth of this increased volume, leaving a final volume of about nine tenths of the original volume in cut.**

Thus for 100 cubic yards of settled embankment 111 cubic yards in cut would be required. But a contractor should have

* See paper by P. J. Flynn in Trans. Tech. Soc. of the Pacific Coast, vol. ii. p. 179, where all the available experimental data are given.

his stakes or poles set one fifth higher than the corresponding fill, so that when filled to the tops of these, a settlement of one sixth will bring the surface to the required grade.

These changes of volume are less for sand and more for stiff, wet clay.

For rock the permanent increase in volume is from 60 to 80 per cent, the greater increase corresponding to a smaller average size of fragment.

326. **Excavations under Water.**—It is often necessary to determine the volume of earth, sand, mud, or rock removed from the beds of rivers, harbors, canals, etc. If this be done by soundings alone, it is likely to work injustice to the contractor, as he would receive no pay for depths excavated below the required limit; and besides, foreign material is apt to flow in and partially replace what is removed, so that the material actually excavated is not adequately shown by soundings within the required limits. It is common, therefore, to pay for the material actually removed, an inspector being usually furnished by the employer to see that no useless work is done beyond the proper bounds. The material is then measured in the dumping scows or barges. The unit of measure is the cubic yard, the same as in earthwork. There are two general methods of gauging scows, or boats. One is to actually measure the inside dimensions of each load, which is often done in the case of rock, and the other is to measure the displacement of the boat, which is the more common method with dredged material. When the barge is gauged by measuring its displacement, the water in the hold must always be pumped down to a given level, or else it must be gauged both before and after loading and the depth of water in the hold observed at each gauging. A displacement diagram (or table) is prepared for each barge, from its actual external dimensions, in terms of its mean draught. There should always be four gaugings taken to determine the draught, at four symmetrically located points

on the sides, these being one fourth the length of the barge from the ends. Fixed gauge-scales, reading to feet and tenths may be painted on the side of the barge, or if it is flat-bottomed, a gauging-rod, with a hook on its lower end at the zero of the scale, may be used and readings taken at these four points. Any distortion of the barge under its load, or any unsymmetrical loading, will then be allowed for, the mean of the four gauge-readings being the true mean draught of the boat.

To prepare a displacement diagram, the areas of the surfaces of displacement must be found for a series of depths uniformly spaced. This series may begin with the depth for no load, the hold being dry. They should then be found for each five tenths of a foot up to the maximum draught. If the boat has plane vertical sides and sloped ends these areas are rectangles, and are readily computed. If the boat is modelled to curved lines, the water-lines can be obtained from the original drawings of the boat, or else they must be obtained by actual measurement. In either case they can be plotted on paper, and their areas determined by a planimeter. These areas are analogous to the cross-sections in the case of railroad earthwork, and the prismoidal formula may be applied for computing the displacement. Thus,

Let A_0, A_1, A_2, A_3, etc., be the areas of the displaced water surfaces, taken at uniform vertical distances h apart. Then for an even number of intervals we have in cubic yards

$$V = \frac{h}{3 \times 27}(A_0 + 4A_1 + 2A_2 + 4A_3 + \dots A_n). \quad (1)$$

If the total range in draught be divided into six equal portions, each equal to h, then Weddel's Rule * would give a

* For the derivation of this rule see Appendix C.

nearer approximation. With the same notation as the above we would then have, in cubic yards,

$$V = \frac{3h}{10}[A_0 + A_2 + A_4 + A_6 + 5(A_1 + A_3 + A_5) + A_3] \ldots \quad (2)$$

These rules are also applicable to the gauging of reservoirs, mill-ponds, or of any irregular volume or cavity.

After the displaced volume of water is found, the corresponding volume of earth or rock is found by applying a proper constant coefficient. This coefficient is always less than unity, and is the reciprocal of the specific gravity of the material. This must be found by experiment. In the case of soft mud it is nearly unity, while with sand and rock it is much more. When rock is purchased by the cubic yard, solid rock is not implied, but the given quality of cut or roughly-quarried rock, piled as closely as possible. When rock is excavated, solid rock is meant. A *measured volume* of any material put into a *gauged scow* will give the proper coefficient for that material. Thus if the measured volume V' give a displacement of V, then $\frac{V'}{V} = C$ is the coefficient to apply to the displacement to give the volume of that material.

TABLES.

TABLE I.

TRIGONOMETRIC FORMULÆ.

TRIGONOMETRIC FUNCTIONS.

Let A (Fig. 107) = angle BAC = arc BF, and let the radius $AF = AB = AH = 1$.

We then have

$\sin A = BC$
$\cos A = AC$
$\tan A = DF$
$\cot A = HG$
$\sec A = AD$
$\operatorname{cosec} A = AG$
$\operatorname{versin} A = CF = BE$
$\operatorname{covers} A = BK = HL$
$\operatorname{exsec} A = BD$
$\operatorname{coexsec} A = BG$
$\operatorname{chord} A = BF$
$\operatorname{chord} 2A = BI = 2BC$

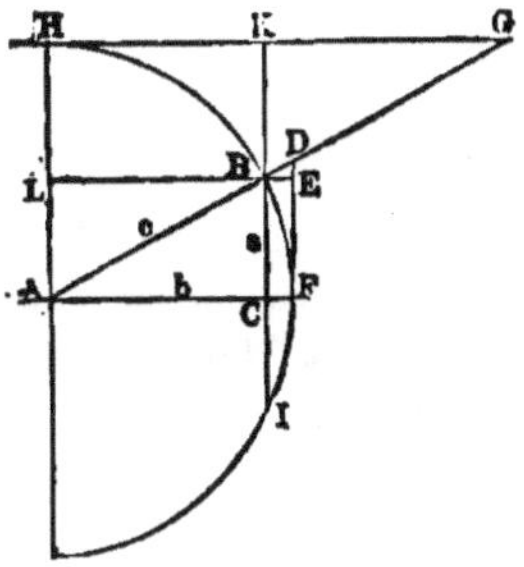

FIG. 107.

In the right-angled triangle ABC (Fig. 107)
Let $AB = c$, $AC = b$, and $BC = a$.
We then have:

1. $\sin A = \frac{a}{c} = \cos B$
2. $\cos A = \frac{b}{c} = \sin B$
3. $\tan A = \frac{a}{b} = \cot B$
4. $\cot A = \frac{b}{a} = \tan B$
5. $\sec A = \frac{c}{b} = \operatorname{cosec} B$
6. $\operatorname{cosec} A = \frac{c}{a} = \sec B$
7. $\operatorname{vers} A = \frac{c-b}{c} = \operatorname{covers} B$
8. $\operatorname{exsec} A = \frac{c-b}{b} = \operatorname{coexsec} B$
9. $\operatorname{covers} A = \frac{c-a}{c} = \operatorname{versin} B$
10. $\operatorname{coexsec} A = \frac{c-a}{a} = \operatorname{exsec} B$
11. $a = c \sin A = b \tan A$
12. $b = c \cos A = a \cot A$
13. $c = \frac{a}{\sin A} = \frac{b}{\cos A}$
14. $a = c \cos B = b \cot B$
15. $b = c \sin B = a \tan B$
16. $c = \frac{a}{\cos B} = \frac{b}{\sin B}$
17. $a = \sqrt{(c+b)(c-b)}$
18. $b = \sqrt{(c+a)(c-a)}$
19. $c = \sqrt{a^2 + b^2}$
20. $C = 90° = A + B$
21. $\text{area} = \frac{ab}{2}$

TABLE I.—*Continued.*

Trigonometric Formulæ.

Solution of Oblique Triangles.

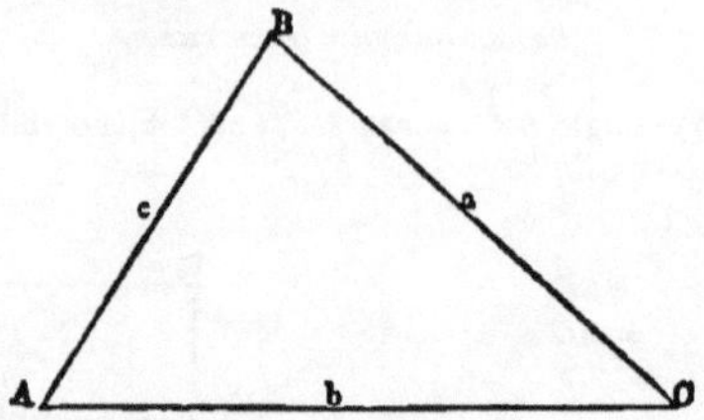

Fig. 108.

	GIVEN.	SOUGHT.	FORMULÆ.
22	A, B, a	C, b, c	$C = 180° - (A + B)$, $b = \frac{a}{\sin A} \cdot \sin B$, $c = \frac{a}{\sin A} \sin (A + B)$
23	A, a, b	B, C, c	$\sin B = \frac{\sin A}{a} \cdot b$, $C = 180° - (A + B)$, $c = \frac{a}{\sin A} \cdot \sin C$.
24	C, a, b	$\frac{1}{2}(A + B)$	$\frac{1}{2}(A + B) = 90° - \frac{1}{2} C$
25		$\frac{1}{2}(A - B)$	$\tan \frac{1}{2}(A - B) = \frac{a - b}{a + b} \tan \frac{1}{2}(A + B)$
26		A, B	$A = \frac{1}{2}(A + B) + \frac{1}{2}(A - B)$, $B = \frac{1}{2}(A + B) - \frac{1}{2}(A - B)$
27		c	$c = (a + b) \frac{\cos \frac{1}{2}(A + B)}{\cos \frac{1}{2}(A - B)} = (a - b) \frac{\sin \frac{1}{2}(A + B)}{\sin \frac{1}{2}(A - B)}$
28		area	$K = \frac{1}{2} a b \sin C$.
29	a, b, c	A	Let $s = \frac{1}{2}(a + b + c)$; $\sin \frac{1}{2} A = \sqrt{\frac{(s-b)(s-c)}{bc}}$
30			$\cos \frac{1}{2} A = \sqrt{\frac{s(s-a)}{bc}}$; $\tan \frac{1}{2} A = \sqrt{\frac{(s-b)(s-c)}{s(s-a)}}$
31			$\sin A = \frac{2\sqrt{s(s-a)(s-b)(s-c)}}{bc}$; $\text{vers } A = \frac{2(s-b)(s-c)}{bc}$
32		area	$K = \sqrt{s(s-a)(s-b)(s-c)}$
33	A, B, C, a	area	$K = \frac{a^2 \sin B \cdot \sin C}{2 \sin A}$

TABLE I.—*Continued.*

TRIGONOMETRIC FORMULÆ.

	GENERAL FORMULÆ.
34	$\sin A = \frac{1}{\operatorname{cosec} A} = \sqrt{1 - \cos^2 A} = \tan A \cos A$
35	$\sin A = 2 \sin \tfrac{1}{2} A \cos \tfrac{1}{2} A = \operatorname{vers} A \cot \tfrac{1}{2} A$
36	$\sin A = \sqrt{\tfrac{1}{2} \operatorname{vers} 2A} = \sqrt{\tfrac{1}{2}(1 - \cos 2A)}$
37	$\cos A = \frac{1}{\sec A} = \sqrt{1 - \sin^2 A} = \cot A \sin A$
38	$\cos A = 1 - \operatorname{vers} A = 2 \cos^2 \tfrac{1}{2} A - 1 = 1 - 2 \sin^2 \tfrac{1}{2} A$
39	$\cos A = \cos^2 \tfrac{1}{2} A - \sin^2 \tfrac{1}{2} A = \sqrt{\tfrac{1}{2} + \tfrac{1}{2} \cos 2A}$
40	$\tan A = \frac{1}{\cot A} = \frac{\sin A}{\cos A} = \sqrt{\sec^2 A - 1}$
41	$\tan A = \sqrt{\frac{1}{\cos^2 A} - 1} = \frac{\sqrt{1 - \cos^2 A}}{\cos A} = \frac{\sin 2A}{1 + \cos 2A}$
42	$\tan A = \frac{1 - \cos 2A}{\sin 2A} = \frac{\operatorname{vers} 2A}{\sin 2A} = \operatorname{exsec} A \cot \tfrac{1}{2} A$
43	$\cot A = \frac{1}{\tan A} = \frac{\cos A}{\sin A} = \sqrt{\operatorname{cosec}^2 A - 1}$
44	$\cot A = \frac{\sin 2A}{1 - \cos 2A} = \frac{\sin 2A}{\operatorname{vers} 2A} = \frac{1 + \cos 2A}{\sin 2A}$
45	$\cot A = \frac{\tan \tfrac{1}{2} A}{\operatorname{exsec} A}$
46	$\operatorname{vers} A = 1 - \cos A = \sin A \tan \tfrac{1}{2} A = 2 \sin^2 \tfrac{1}{2} A$
47	$\operatorname{vers} A = \operatorname{exsec} A \cos A$
48	$\operatorname{exsec} A = \sec A - 1 = \tan A \tan \tfrac{1}{2} A = \frac{\operatorname{vers} A}{\cos A}$
49	$\sin \tfrac{1}{2} A = \sqrt{\frac{1 - \cos A}{2}} = \sqrt{\frac{\operatorname{vers} A}{2}}$
50	$\sin 2A = 2 \sin A \cos A$
51	$\cos \tfrac{1}{2} A = \sqrt{\frac{1 + \cos A}{2}}$
52	$\cos 2A = 2 \cos^2 A - 1 = \cos^2 A - \sin^2 A = 1 - 2 \sin^2 A$

TABLE I.—*Continued.*

Trigonometric Formulæ.

General Formulæ.

53. $$\tan \tfrac{1}{2} A = \frac{\tan A}{1 + \sec A} = \operatorname{cosec} A - \cot A = \frac{1 - \cos A}{\sin A} = \sqrt{\frac{1 - \cos A}{1 + \cos A}}$$

54. $$\tan 2A = \frac{2 \tan A}{1 - \tan^2 A}$$

55. $$\cot. \tfrac{1}{2} A = \frac{\sin A}{\operatorname{vers} A} = \frac{1 + \cos A}{\sin A} = \frac{1}{\operatorname{cosec} A - \cot A}$$

56. $$\cot 2A = \frac{\cot^2 A - 1}{2 \cot A}$$

57. $$\operatorname{vers} \tfrac{1}{2} A = \frac{\tfrac{1}{2} \operatorname{vers} A}{1 + \sqrt{1 - \tfrac{1}{2} \operatorname{vers} A}} = \frac{1 - \cos A}{2 + \sqrt{2(1 + \cos A)}}$$

58. $$\operatorname{vers} 2A = 2 \sin^2 A$$

59. $$\operatorname{exsec} \tfrac{1}{2} A = \frac{1 - \cos A}{(1 + \cos A) + \sqrt{2(1 + \cos A)}}$$

60. $$\operatorname{exsec} 2A = \frac{\tan^2 A}{1 - \tan^2 A}$$

61. $$\sin (A \pm B) = \sin A . \cos B \pm \sin B . \cos A$$

62. $$\cos (A \pm B) = \cos A . \cos B \mp \sin A . \sin B$$

63. $$\sin A + \sin B = 2 \sin \tfrac{1}{2}(A + B) \cos \tfrac{1}{2}(A - B)$$

64. $$\sin A - \sin B = 2 \cos \tfrac{1}{2}(A + B) \sin \tfrac{1}{2}(A - B)$$

65. $$\cos A + \cos B = 2 \cos \tfrac{1}{2}(A + B) \cos \tfrac{1}{2}(A - B)$$

66. $$\cos B - \cos A = 2 \sin \tfrac{1}{2}(A + B) \sin \tfrac{1}{2}(A - B)$$

67. $$\sin^2 A - \sin^2 B = \cos^2 B - \cos^2 A = \sin (A + B) \sin (A - B)$$

68. $$\cos^2 A - \sin^2 B = \cos (A + B) \cos (A - B)$$

69. $$\tan A + \tan B = \frac{\sin (A + B)}{\cos A . \cos B}$$

70. $$\tan A - \tan B = \frac{\sin (A - B)}{\cos A . \cos B}$$

TABLE II.

For Converting Metres, Feet, and Chains.

Metres to Feet.		Feet to Metres and Chains.			Chains to Feet.	
Metres.	Feet.	Feet.	Metres.	Chains.	Chains.	Feet.
1	3.28087	1	0.304797	0.0151	0.01	0.66
2	6.56174	2	0.609595	.0303	.02	1.32
3	9.84261	3	0.914392	.0455	.03	1.98
4	13.12348	4	1.219189	.0606	.04	2.64
5	16.40435	5	1.523986	.0758	.05	3.30
6	19.68522	6	1.828784	.0909	.06	3.96
7	22.96609	7	2.133581	.1061	.07	4.62
8	26.24695	8	2.438378	.1212	.08	5.28
9	29.52782	9	2.743175	.1364	.09	5.94
10	32.80869	10	3.047973	.1515	.10	6.60
20	65.61739	20	6.095946	.3030	.20	13.20
30	98.42609	30	9.143918	.4545	.30	19.80
40	131.2348	40	12.19189	.6061	.40	26.40
50	164.0435	50	15.23986	.7576	.50	33.00
60	196.8522	60	18.28784	.9091	.60	39.60
70	229.6609	70	21.33581	1.0606	.70	46.20
80	262.4695	80	24.38378	1.2121	.80	52.80
90	295.2782	90	27.43175	1.3636	.90	59.40
100	328.0869	100	30.47973	1.5151	1	66.00
200	656.1739	100	60.95946	3.0303	2	132
300	984.2609	300	91.43918	4.5455	3	198
400	1312.348	400	121.9189	6.0606	4	264
500	1640.435	500	152.3986	7.5756	5	330
600	1968.522	600	182.8784	9.0909	6	396
700	2296.609	700	213.3581	10.606	7	462
800	2624.695	800	243.8378	12.121	8	528
900	2952.782	900	274.3175	13.636	9	594
1000	3280.869	1000	304.7973	15.151	10	660
2000	6561.739	2000	609.5946	30.303	20	1320
3000	9842.609	3000	914.3918	45.455	30	1980
4000	13123.48	4000	1219.189	60.606	40	2640
5000	16404.35	5000	1523.986	75.756	50	3300
6000	19685.22	6000	1828.784	90.909	60	3960
7000	22966.09	7000	2133.581	106.06	70	4620
8000	26246.95	8000	2438.378	121.21	80	5280
9000	29527.82	9000	2743.175	136.36	90	5940

TABLE III.

Logarithms of Numbers. § 173.

Nat. Nos.	0	1	2	3	4	5	6	7	8	9	Proportional Parts.								
											1	2	3	4	5	6	7	8	9
10	.0000	.0043	.0086	.0128	.0170	.0212	.0253	.0294	.0334	.0374	4	8	12	17	21	25	29	33	37
11	.0414	.0453	.0492	.0531	.0569	.0607	.0645	.0682	.0719	.0755	4	8	11	15	19	23	26	30	34
12	.0792	.0828	.0864	.0899	.0934	.0969	.1004	.1038	.1072	.1106	3	7	10	14	17	21	24	28	31
13	.1139	.1173	.1206	.1239	.1271	.1303	.1335	.1367	.1399	.1430	3	6	10	13	16	19	23	26	29
14	.1461	.1492	.1523	.1553	.1584	.1614	.1644	.1673	.1703	.1732	3	6	9	12	15	18	21	24	27
15	.1761	.1790	.1818	.1847	.1875	.1903	.1931	.1959	.1987	.2014	3	6	8	11	14	17	20	22	25
16	.2041	.2068	.2095	.2122	.2148	.2175	.2201	.2227	.2253	.2279	3	5	8	11	13	16	18	21	24
17	.2304	.2330	.2355	.2380	.2405	.2430	.2455	.2480	.2504	.2529	2	5	7	10	12	15	17	20	22
18	.2553	.2577	.2601	.2625	.2648	.2672	.2695	.2718	.2742	.2765	2	5	7	9	12	14	16	19	21
19	.2788	.2810	.2833	.2856	.2878	.2900	.2923	.2945	.2967	.2989	2	4	7	9	11	13	16	18	20
20	.3010	.3032	.3054	.3075	.3096	.3118	.3139	.3160	.3181	.3201	2	4	6	8	11	13	15	17	19
21	.3222	.3243	.3263	.3284	.3304	.3324	.3345	.3365	.3385	.3404	2	4	6	8	10	12	14	16	18
22	.3424	.3444	.3464	.3483	.3502	.3522	.3541	.3560	.3579	.3598	2	4	6	8	10	12	14	15	17
23	.3617	.3636	.3655	.3674	.3692	.3711	.3729	.3747	.3766	.3784	2	4	6	7	9	11	13	15	17
24	.3802	.3820	.3838	.3856	.3874	.3892	.3909	.3927	.3945	.3962	2	4	5	7	9	11	12	14	16
25	.3979	.3997	.4014	.4031	.4048	.4065	.4082	.4099	.4116	.4133	2	3	5	7	9	10	12	14	15
26	.4150	.4166	.4183	.4200	.4216	.4232	.4249	.4265	.4281	.4298	2	3	5	7	8	10	11	13	15
27	.4314	.4330	.4346	.4362	.4378	.4393	.4409	.4425	.4440	.4456	2	3	5	6	8	9	11	13	14
28	.4472	.4487	.4502	.4518	.4533	.4548	.4564	.4579	.4594	.4609	2	3	5	6	8	9	11	12	14
29	.4624	.4639	.4654	.4669	.4683	.4698	.4713	.4728	.4742	.4757	1	3	4	6	7	9	10	12	13
30	.4771	.4786	.4800	.4814	.4829	.4843	.4857	.4871	.4886	.4900	1	3	4	6	7	9	10	11	13
31	.4914	.4928	.4942	.4955	.4969	.4983	.4997	.5011	.5024	.5038	1	3	4	6	7	8	10	11	12
32	.5051	.5065	.5079	.5092	.5105	.5119	.5132	.5145	.5159	.5172	1	3	4	5	7	8	9	11	12
33	.5185	.5198	.5211	.5224	.5237	.5250	.5263	.5276	.5289	.5302	1	3	4	5	6	8	9	10	12
34	.5315	.5328	.5340	.5353	.5366	.5378	.5391	.5403	.5416	.5428	1	3	4	5	6	8	9	10	11
35	.5441	.5453	.5465	.5478	.5490	.5502	.5514	.5527	.5539	.5551	1	2	4	5	6	7	9	10	11
36	.5563	.5575	.5587	.5599	.5611	.5623	.5635	.5647	.5658	.5670	1	2	4	5	6	7	8	10	11
37	.5682	.5694	.5705	.5717	.5729	.5740	.5752	.5763	.5775	.5786	1	2	3	5	6	7	8	9	10
38	.5798	.5809	.5821	.5832	.5843	.5855	.5866	.5877	.5888	.5899	1	2	3	5	6	7	8	9	10
39	.5911	.5922	.5933	.5944	.5955	.5966	.5977	.5988	.5999	.6010	1	2	3	4	5	7	8	9	10
40	.6021	.6031	.6042	.6053	.6064	.6075	.6085	.6096	.6107	.6117	1	2	3	4	5	6	8	9	10
41	.6128	.6138	.6149	.6160	.6170	.6180	.6191	.6201	.6212	.6222	1	2	3	4	5	6	7	8	9
42	.6232	.6243	.6253	.6263	.6274	.6284	.6294	.6304	.6314	.6325	1	2	3	4	5	6	7	8	9
43	.6335	.6345	.6355	.6365	.6375	.6385	.6395	.6405	.6415	.6425	1	2	3	4	5	6	7	8	9
44	.6435	.6444	.6454	.6464	.6474	.6484	.6493	.6503	.6513	.6522	1	2	3	4	5	6	7	8	9
45	.6532	.6542	.6551	.6561	.6571	.6580	.6590	.6599	.6609	.6618	1	2	3	4	5	6	7	8	9
46	.6628	.6637	.6646	.6656	.6665	.6675	.6684	.6693	.6702	.6712	1	2	3	4	5	6	7	7	8
47	.6721	.6730	.6739	.6749	.6758	.6767	.6776	.6785	.6794	.6803	1	2	3	4	5	5	6	7	8
48	.6812	.6821	.6830	.6839	.6848	.6857	.6866	.6875	.6884	.6893	1	2	3	4	4	5	6	7	8
49	.6902	.6911	.6920	.6928	.6937	.6946	.6955	.6964	.6972	.6981	1	2	3	4	4	5	6	7	8
50	.6990	.6998	.7007	.7016	.7024	.7033	.7042	.7050	.7059	.7067	1	2	3	3	4	5	6	7	8
51	.7076	.7084	.7093	.7101	.7110	.7118	.7126	.7135	.7143	.7152	1	2	3	3	4	5	6	7	8
52	.7160	.7168	.7177	.7185	.7193	.7202	.7210	.7218	.7226	.7235	1	2	2	3	4	5	6	7	7
53	.7243	.7251	.7259	.7267	.7275	.7284	.7292	.7300	.7308	.7316	1	2	2	3	4	5	6	6	7
54	.7324	.7332	.7340	.7348	.7356	.7364	.7372	.7380	.7388	.7396	1	2	2	3	4	5	6	6	7

TABLE III.—*Continued.*

Logarithms of Numbers.

Nat. Nos.	0	1	2	3	4	5	6	7	8	9	Proportional Parts. 1	2	3	4	5	6	7	8	9
55	.7404	.7412	.7419	.7427	.7435	.7443	.7451	.7459	.7466	.7474	1	2	2	3	4	5	5	6	7
56	.7482	.7490	.7497	.7505	.7513	.7520	.7528	.7536	.7543	.7551	1	2	2	3	4	5	5	6	7
57	.7559	.7566	.7574	.7582	.7589	.7597	.7604	.7612	.7619	.7627	1	2	2	3	4	5	5	6	7
58	.7634	.7642	.7649	.7657	.7664	.7672	.7679	.7686	.7694	7701	1	1	2	3	4	4	5	6	7
59	.7709	.7716	.7723	.7731	.7738	.7745	.7752	.7760	.7767	.7774	1	1	2	3	4	4	5	6	7
60	.7782	.7789	.7796	.7803	.7810	.7818	.7825	.7832	.7839	.7846	1	1	2	3	4	4	5	6	6
61	.7853	.7860	.7868	.7875	.7882	.7889	.7896	.7903	.7910	.7917	1	1	2	3	4	4	5	6	6
62	.7924	.7931	.7938	.7945	.7952	.7959	.7966	.7973	.7980	.7987	1	1	2	3	3	4	5	6	6
63	.7993	.8000	.8007	.8014	.8021	.8028	.8035	.8041	.8048	.8055	1	1	2	3	3	4	5	5	6
64	.8062	.8069	.8075	.8082	.8089	.8096	.8102	.8109	.8116	.8122	1	1	2	3	3	4	5	5	6
65	.8129	.8136	.8142	.8149	.8156	.8162	.8169	.8176	.8182	.8189	1	1	2	3	3	4	5	5	6
66	.8195	.8202	.8209	.8215	.8222	.8228	.8235	.8241	.8248	.8254	1	1	2	3	3	4	5	5	6
67	.8261	.8267	.8274	.8280	.8287	.8293	.8299	.8306	.8312	.8319	1	1	2	3	3	4	5	5	6
68	.8325	.8331	.8338	.8344	.8351	.8357	.8363	.8370	.8376	.8382	1	1	2	3	3	4	4	5	6
69	.8388	.8395	.8401	.8407	.8414	.8420	.8426	.8432	.8439	.8445	1	1	2	2	3	4	4	5	6
70	.8451	.8457	.8463	.8470	.8476	.8482	.8488	.8494	.8500	.8506	1	1	2	2	3	4	4	5	6
71	.8513	.8519	.8525	.8531	.8537	.8543	.8549	.8555	.8561	.8567	1	1	2	2	3	4	4	5	5
72	.8573	.8579	.8585	.8591	.8597	.8603	.8609	.8615	.8621	.8627	1	1	2	2	3	4	4	5	5
73	.8633	.8639	.8645	.8651	.8657	.8663	.8669	.8675	.8681	.8686	1	1	2	2	3	4	4	5	5
74	.8692	.8698	.8704	.8710	.8716	.8722	.8727	.8733	.8739	.8745	1	1	2	2	3	4	4	5	5
75	.8751	.8756	.8762	.8768	.8774	.8779	.8785	.8791	.8797	.8802	1	1	2	2	3	3	4	5	5
76	.8808	.8814	.8820	.8825	.8831	.8837	.8842	.8848	.8854	.8859	1	1	2	2	3	3	4	5	5
77	.8865	.8871	.8876	.8882	.8887	.8893	.8899	.8904	.8910	.8915	1	1	2	2	3	3	4	4	5
78	.8921	.8927	.8932	.8938	.8943	.8949	.8954	.8960	.8965	.8971	1	1	2	2	3	3	4	4	5
79	.8976	.8982	.8987	.8993	.8998	.9004	.9009	.9015	.9020	.9025	1	1	2	2	3	3	4	4	5
80	.9031	.9036	.9042	.9047	.9053	.9058	.9063	.9069	.9074	.9079	1	1	2	2	3	3	4	4	5
81	.9085	.9090	.9096	.9101	.9106	.9112	.9117	.9122	.9128	.9133	1	1	2	2	3	3	4	4	5
82	.9138	.9143	.9149	.9154	.9159	.9165	.9170	.9175	.9180	.9186	1	1	2	2	3	3	4	4	5
83	.9191	.9196	.9201	.9206	.9212	.9217	.9222	.9227	.9232	.9238	1	1	2	2	3	3	4	4	5
84	.9243	.9248	.9253	.9258	.9263	.9269	.9274	.9279	.9284	.9289	1	1	2	2	3	3	4	4	5
85	.9294	.9299	.9304	.9309	.9315	.9320	.9325	.9330	.9335	.9340	1	1	2	2	3	3	4	4	5
86	.9345	.9350	.9355	.9360	.9365	.9370	.9375	.9380	.9385	.9390	1	1	2	2	3	3	4	4	5
87	.9395	.9400	.9405	.9410	.9415	.9420	.9425	.9430	.9435	.9440	0	1	1	2	2	3	3	4	4
88	.9445	.9450	.9455	.9460	.9465	.9469	.9474	.9479	.9484	.9489	0	1	1	2	2	3	3	4	4
89	.9494	.9499	.9504	.9509	.9513	.9518	.9523	.9528	.9533	.9538	0	1	1	2	2	3	3	4	4
90	.9542	.9547	.9552	.9557	.9562	.9566	.9571	.9576	.9581	.9586	0	1	1	2	2	3	3	4	4
91	.9590	.9595	.9600	.9605	.9609	.9614	.9619	.9624	.9628	.9633	0	1	1	2	2	3	3	4	4
92	.9638	.9643	.9647	.9652	.9657	.9661	.9666	.9671	.9675	.9680	0	1	1	2	2	3	3	4	4
93	.9685	.9689	.9694	.9699	.9703	.9708	.9713	.9717	.9722	.9727	0	1	1	2	2	3	3	4	4
94	.9731	.9736	.9741	.9745	.9750	.9754	.9759	.9763	.9768	.9773	0	1	1	2	2	3	3	4	4
95	.9777	.9782	.9786	.9791	.9795	.9800	.9805	.9809	.9814	.9818	0	1	1	2	2	3	3	4	4
96	.9823	.9827	.9832	.9836	.9841	.9845	.9850	.9854	.9859	.9863	0	1	1	2	2	3	3	4	4
97	.9868	.9872	.9877	.9881	.9886	.9890	.9894	.9899	.9903	.9908	0	1	1	2	2	3	3	4	4
98	.9912	.9917	.9921	.9926	.9930	.9934	.9939	.9943	.9948	.9952	0	1	1	2	2	3	3	4	4
99	.9956	.9961	.9965	.9969	.9974	.9978	.9983	.9987	.9991	.9996	0	1	1	2	2	3	3	3	4

TABLE IIIA.
LOGARITHMS OF SINES AND TANGENTS.

	0°				1°				
	Sin.	Cos.	Tan.	Cot.	Sin.	Cos.	Tan.	Cot.	
0′		0.0000			8.2419	9.9999	8.2419	1.7581	60′
1	6.4637	.0000	6.4637	3.5363	.2490	.9999	.2491	.7509	59
2	.7648	.0000	.7648	.2352	.2561	.9999	.2562	.7438	58
3	6.9408	.0000	6.9408	3.0592	.2630	.9999	.2631	.7369	57
4	7.0658	.0000	7.0658	2.9342	.2699	.9999	.2700	.7300	56
5	.1627	.0000	.1627	.8373	.2766	.9999	.2767	.7233	55
6	.2419	.0000	.2419	.7581	.2832	.9999	.2833	.7167	54
7	.3088	.0000	.3088	.6912	.2898	.9999	.2899	.7101	53
8	.3668	.0000	.3668	.6332	.2962	.9999	.2963	.7037	52
9	.4180	.0000	.4180	.5820	.3025	.9999	.3026	.6974	51
10	.4637	.0000	.4637	.5363	.3088	.9999	.3089	.6911	50
11	.5051	.0000	.5051	.4949	.3150	.9999	.3150	.6850	49
12	.5429	.0000	.5429	.4571	.3210	.9999	.3211	.6789	48
13	.5777	.0000	.5777	.4223	.3270	.9999	.3271	.6729	47
14	.6099	.0000	.6099	.3901	.3329	.9999	.3330	.6670	46
15	.6398	.0000	.6398	.3602	.3388	.9999	.3389	.6611	45
16	.6678	.0000	.6678	.3322	.3445	.9999	.3446	.6554	44
17	.6942	.0000	.6942	.3058	.3502	.9999	.3503	.6497	43
18	.7190	.0000	.7190	.2810	.3558	.9999	.3559	.6441	42
19	.7425	.0000	.7425	.2575	.3613	.9999	.3614	.6386	41
20	.7648	.0000	.7648	.2352	.3668	.9999	.3669	.6331	40
21	.7859	.0000	.7860	.2140	.3722	.9999	.3723	.6277	39
22	.8061	.0000	.8062	.1938	.3775	.9999	.3776	.6224	38
23	.8255	.0000	.8255	.1745	.3828	.9999	.3829	.6171	37
24	.8439	.0000	.8439	.1561	.3880	.9999	.3881	.6119	36
25	.8617	.0000	.8617	.1383	.3931	.9999	.3932	.6068	35
26	.8787	.0000	.8787	.1213	.3982	.9999	.3983	.6017	34
27	.8951	.0000	.8951	.1049	.4032	.9999	.4033	.5967	33
28	.9109	.0000	.9109	.0891	.4082	.9999	.4083	.5917	32
29	.9261	.0000	.9261	.0739	.4131	.9999	.4132	.5868	31
30	.9408	.0000	.9409	.0591	.4179	.9999	.4181	.5819	30
31	.9551	.0000	.9551	.0449	.4227	.9998	.4229	.5771	29
32	.9689	.0000	.9689	.0311	.4275	.9998	.4276	.5724	28
33	.9822	.0000	.9823	.0177	.4322	.9998	.4323	.5677	27
34	7.9952	.0000	7.9952	2.0048	.4368	.9998	.4370	.5630	26
35	8.0078	.0000	8.0078	1.9922	.4414	.9998	.4416	.5584	25
36	.0200	.0000	.0200	.9800	.4459	.9998	.4461	.5539	24
37	.0319	.0000	.0319	.9681	.4504	.9998	.4506	.5494	23
38	.0435	.0000	.0435	.9565	.4549	.9998	.4551	.5449	22
39	.0548	.0000	.0548	.9452	.4593	.9998	.4595	.5405	21
40	.0658	.0000	.0658	.9342	.4637	.9998	.4638	.5362	20
41	.0765	.0000	.0765	.9235	.4680	.9998	.4682	.5318	19
42	.0870	.0000	.0870	.9130	.4723	.9998	.4725	.5275	18
43	.0972	.0000	.0972	.9028	.4765	.9998	.4767	.5233	17
44	.1072	.0000	.1072	.8928	.4807	.9998	.4809	.5191	16
45	.1169	.0000	.1170	.8830	.4848	.9998	.4851	.5149	15
46	.1265	.0000	.1265	.8735	.4890	.9998	.4892	.5108	14
47	.1358	.0000	.1359	.8641	.4930	.9998	.4933	.5067	13
48	.1450	.0000	.1450	.8550	.4971	.9998	.4973	.5027	12
49	.1539	.0000	.1540	.8460	.5011	.9998	.5013	.4987	11
50	.1627	.0000	.1627	.8373	.5050	.9998	.5053	.4947	10
51	.1713	.0000	.1713	.8287	.5090	.9998	.5092	.4908	9
52	.1797	0.0000	.1798	.8202	.5129	.9998	.5131	.4869	8
53	.1880	9.9999	.1880	.8120	.5167	.9998	.5170	.4830	7
54	.1961	.9999	.1962	.8038	.5206	.9998	.5208	.4792	6
55	.2041	.9999	.2041	.7959	.5243	.9998	.5246	.4754	5
56	.2119	.9999	.2120	.7880	.5281	.9998	.5283	.4717	4
57	.2196	.9999	.2196	.7804	.5318	.9997	.5321	.4679	3
58	.2271	.9999	.2272	.7728	.5355	.9997	.5358	.4642	2
59	.2346	.9999	.2346	.7654	.5392	.9997	.5394	.4606	1
60	8.2419	9.9999	8.2419	1.7581	8.5428	9.9997	8.5431	1.4569	0
	Cos.	Sin.	Cot.	Tan.	Cos.	Sin.	Cot.	Tan.	
	89°				88°				

TABLE IIIa.—*Continued.*

Logarithms of Sines and Tangents.

	2°				3°				4°				
	Sin.	Cos.	Tan.	Cot.	Sin.	Cos.	Tan.	Cot.	Sin.	Cos.	Tan.	Cot.	
0′	8.5428	9.9997	8.5431	1.4569	8.7188	9.9994	8.7194	1.2806	8.8436	9.9989	8.8446	1.1554	60′
1	.5464	.9997	.5467	.4533	.7212	.9994	.7218	.2782	.8454	.9989	.8465	.1535	59
2	.5500	.9997	.5503	.4497	.7236	.9994	.7242	.2758	.8472	.9989	.8483	.1517	58
3	.5535	.9997	.5538	.4462	.7260	.9994	.7266	.2734	.8490	.9989	.8501	.1499	57
4	.5571	.9997	.5573	.4427	.7283	.9994	.7290	.2710	.8508	.9989	.8518	.1482	56
5	.5605	.9997	.5608	.4392	.7307	.9994	.7313	.2687	.8525	.9989	.8536	.1464	55
6	.5640	.9997	.5643	.4357	.7330	.9994	.7337	.2663	.8543	.9989	.8554	.1446	54
7	.5674	.9997	.5677	.4323	.7354	.9994	.7360	.2640	.8560	.9989	.8572	.1428	53
8	.5708	.9997	.5711	.4289	.7377	.9994	.7383	.2617	.8578	.9989	.8589	.1411	52
9	.5742	.9997	.5745	.4255	.7400	.9993	.7406	.2594	.8595	.9989	.8607	.1393	51
10	.5776	.9997	.5779	.4221	.7423	.9993	.7429	.2571	.8613	.9989	.8624	.1376	50
11	.5809	.9997	.5812	.4188	.7445	.9993	.7452	.2548	.8630	.9988	.8642	.1358	49
12	.5842	.9997	.5845	.4155	.7468	.9993	.7475	.2525	.8647	.9988	.8659	.1341	48
13	.5875	.9997	.5878	.4122	.7491	.9993	.7497	.2503	.8665	.9988	.8676	.1324	47
14	.5907	.9997	.5911	.4089	.7513	.9993	.7520	.2480	.8682	.9988	.8694	.1306	46
15	.5939	.9997	.5943	.4057	.7535	.9993	.7542	.2458	.8699	.9988	.8711	.1289	45
16	.5972	.9997	.5975	.4025	.7557	.9993	.7565	.2435	.8716	.9988	.8728	.1272	44
17	.6003	.9997	.6007	.3993	.7580	.9993	.7587	.2413	.8733	.9988	.8745	.1255	43
18	.6035	.9996	.6038	.3962	.7602	.9993	.7609	.2391	.8749	.9988	.8762	.1238	42
19	.6066	.9996	.6070	.3930	.7623	.9993	.7631	.2369	.8766	.9988	.8778	.1222	41
20	.6097	.9996	.6101	.3899	.7645	.9993	.7652	.2348	.8783	.9988	.8795	.1205	40
21	.6128	.9996	.6132	.3868	.7667	.9993	.7674	.2326	.8799	.9987	.8812	.1188	39
22	.6159	.9996	.6163	.3837	.7688	.9992	.7696	.2304	.8816	.9987	.8829	.1171	38
23	.6189	.9996	.6193	.3807	.7710	.9992	.7717	.2283	.8833	.9987	.8845	.1155	37
24	.6220	.9996	.6223	.3777	.7731	.9992	.7739	.2261	.8849	.9987	.8862	.1138	36
25	.6250	.9996	.6254	.3746	.7752	.9992	.7760	.2240	.8865	.9987	.8878	.1122	35
26	.6279	.9996	.6283	.3717	.7773	.9992	.7781	.2219	.8882	.9987	.8895	.1105	34
27	.6309	.9996	.6313	.3687	.7794	.9992	.7802	.2198	.8898	.9987	.8911	.1089	33
28	.6339	.9996	.6343	.3657	.7815	.9992	.7823	.2177	.8914	9987	.8927	.1073	32
29	.6368	.9996	.6372	.3628	.7836	.9992	.7844	.2156	.8930	.9987	.8944	.1056	31
30	.6397	.9996	.6401	.3599	.7857	.9992	.7865	.2135	.8946	.9987	.8960	.1040	30
31	.6426	.9996	.6430	.3570	.7877	.9992	.7886	.2114	.8962	.9986	.8976	.1024	29
32	.6454	.9996	.6459	.3541	.7898	.9992	.7906	.2094	.8978	.9986	.8992	.1008	28
33	.6483	.9996	.6487	.3513	.7918	.9992	.7927	.2073	.8994	.9986	.9008	.0992	27
34	.6511	.9996	.6515	.3485	.7939	.9992	.7947	.2053	.9010	.9986	.9024	.0976	26
35	.6539	.9996	.6544	.3456	.7959	.9992	.7967	.2033	.9026	.9986	.9040	.0960	25
36	.6567	.9996	.6571	.3429	.7979	.9991	.7988	.2012	.9342	.9986	.9056	.0944	24
37	.6595	.9995	.6599	.3401	.7999	.9991	.8008	.1992	.9057	.9986	.9071	.0929	23
38	.6622	.9995	.6627	.3373	.8019	.9991	.8028	.1972	.9073	.9986	.9087	.0913	22
39	.6650	.9995	.6654	.3346	.8039	.9991	.8048	.1952	.9089	.9986	.9103	.0897	21
40	.6677	.9995	.6682	.3318	.8059	.9991	.8067	.1933	.9104	.9986	.9118	.0882	20
41	.6704	.9995	.6709	.3291	.8078	.9991	.8087	.1913	.9119	.9985	.9134	.0866	19
42	.6731	.9995	.6736	.3264	.8098	.9991	.8107	.1893	.9135	.9985	.9150	.0850	18
43	.6758	.9995	.6762	.3238	.8117	.9991	.8126	.1874	.9150	.9985	.9165	.0835	17
44	.6784	.9995	.6789	.3211	.8137	.9991	.8146	.1854	.9166	.9985	.9180	.0820	16
45	.6810	.9995	.6815	.3185	.8156	.9991	.8165	.1835	.9181	.9985	.9196	.0804	15
46	.6837	.9995	.6842	.3158	.8175	.9991	.8185	.1815	.9196	.9985	.9211	.0789	14
47	.6863	.9995	.6868	.3132	.8194	.9991	.8204	.1796	.9211	.9985	.9226	.0774	13
48	.6889	.9995	.6894	.3106	.8213	.9990	.8223	.1777	.9226	.9985	.9241	.0759	12
49	.6914	.9995	.6920	.3080	.8232	.9990	.8242	.1758	.9241	.9985	.9256	.0744	11
50	.6940	.9995	.6945	.3055	.8251	.9990	.8261	.1739	.9256	.9985	.9272	.0728	10
51	.6965	.9995	.6971	.3029	.8270	.9990	.8280	.1720	.9271	.9984	.9287	.0713	9
52	.6991	.9995	.6996	.3004	.8289	.9990	.8299	.1701	.9286	.9984	.9302	.0698	8
53	.7016	.9994	.7021	.2979	.8307	.9990	.8317	.1683	.9301	.9984	.9316	.0684	7
54	.7041	.9994	.7046	.2954	.8326	.9990	.8336	.1664	.9315	.9984	.9331	.0669	6
55	.7066	.9994	.7071	.2929	.8345	.9990	.8355	.1645	.9330	.9984	.9346	.0654	5
56	.7090	.9994	.7096	.2904	.8363	.9990	.8373	.1627	.9345	.9984	.9361	.0639	4
57	.7115	.9994	.7121	.2879	.8381	.9990	.8392	.1608	.9359	.9984	.9376	.0624	3
58	.7140	.9994	.7145	.2855	.8400	.9990	.8410	.1590	.9374	.9984	.9390	.0610	2
59	.7164	.9994	.7170	.2830	.8418	.9989	.8428	.1572	.9388	.9984	.9405	.0595	1
60	8.7188	9.9994	8.7194	1.2806	8 8436	9.9989	8.8446	1.1554	8 9403	9.9983	8.9420	1.0580	0
	Cos.	Sin.	Cot.	Tan.	Cos.	Sin.	Cot.	Tan.	Cos.	Sin.	Cot.	Tan.	
	87°				86°				85°				

TABLE IIIA—*Continued.*

Logarithms of Sines and Tangents.

Arc.	Sin.	Df.	Cos.	Df.	Tan.	Df.	Cot.	Arc.	Arc.	Sin.	Df.	Cos.	Df.	Tan.	Df.	Cot.	Arc.
° ′								° ′	° ′								° ′
5 0	8.9403	142	9.9983	1	8.9420	143	1.0580	85 0	15 0	9.4130	47	9.9849	3	9.4281	50	0.5719	75 0
10	.9545	137	.9982	1	.9563	138	.0437	50	10	.4177	46	.9846	3	.4331	50	.5669	50
20	.9682	124	.9981	1	.9701	135	.0299	40	20	.4223	46	.9843	4	.4381	49	.5619	40
30	.9816	129	.9980	1	.9836	130	.0164	30	30	.4269	45	.9839	3	.4430	49	.5570	30
40	8.9945	125	.9979	2	8.9966	127	1.0034	20	40	.4314	45	.9836	4	.4479	48	.5521	20
50	9.0070	122	.9977	1	9.0093	123	0.9907	10	50	.4359	44	.9832	4	.4527	48	.5473	10
6 0	.0192	119	.9976	1	.0216	120	.9784	84 0	16 0	.4403	44	.9828	3	.4575	47	.5425	74 0
10	.0311	115	.9975	2	.0336	117	.9664	50	10	.4447	44	.9825	4	.4622	47	.5378	50
20	.0426	113	.9973	1	.0453	114	.9547	40	20	.4491	42	.9821	4	.4669	47	.5331	40
30	.0539	109	.9972	1	.0567	111	.9433	30	30	.4533	43	.9817	3	.4716	46	.5284	30
40	.0648	107	.9971	2	.0678	108	.9322	20	40	.4576	42	.9814	4	.4762	46	.5238	20
50	.0755	104	.9969	1	.0786	105	.9214	10	50	.4618	41	.9810	4	.4808	45	.5192	10
7 0	.0859	102	.9968	2	.0891	104	.9109	83 0	17 0	.4659	41	.9806	4	.4853	45	.5147	73 0
10	.0961	99	.9966	2	.0995	101	.9005	50	10	.4700	41	.9802	4	.4898	45	.5102	50
20	.1060	97	.9964	1	.1096	98	.8904	40	20	.4741	40	.9798	4	.4943	44	.5057	40
30	.1157	95	.9963	2	.1194	97	.8806	30	30	.4781	40	.9794	4	.4987	44	.5013	30
40	.1252	93	.9961	2	.1291	94	.8709	20	40	.4821	40	.9790	4	.5031	44	.4969	20
50	.1345	91	.9959	1	.1385	93	.8615	10	50	.4861	39	.9786	4	.5075	43	.4925	10
8 0	.1436	89	.9958	2	.1478	91	.8522	82 0	18 0	.4900	39	.9782	4	.5118	43	.4882	72 0
10	.1525	87	.9956	2	.1569	89	.8431	50	10	.4939	38	.9778	4	.5161	42	.4839	50
20	.1612	85	.9954	2	.1658	87	.8342	40	20	.4977	38	.9774	4	.5203	42	.4797	40
30	.1697	84	.9952	2	.1745	86	.8255	30	30	.5015	37	.9770	5	.5245	42	.4755	30
40	.1781	82	.9950	2	.1831	84	.8169	20	40	.5052	38	.9765	4	.5287	42	.4713	20
50	.1863	80	.9948	2	.1915	82	.8085	10	50	.5090	36	.9761	4	.5329	41	.4671	10
9 0	.1913	79	.9946	2	.1997	81	.8003	81 0	19 0	.5126	37	.9757	5	.5370	41	.4630	71 0
10	.2022	78	.9944	2	.20[illegible]	80	.7922	50	10	.5163	36	.9752	4	.5411	40	.4589	50
20	.2100	76	.9942	2	.2158	78	.7842	40	20	.5199	36	.9748	5	.5451	40	.4549	40
30	.2176	75	.9940	2	.2236	77	.7764	30	30	.5235	35	.9743	4	.5491	40	.4509	30
40	.2251	73	.9938	2	.2313	76	.7687	20	40	.5270	36	.9739	5	.5531	40	.4469	20
50	.2324	73	.9936	2	.2389	74	.7611	10	50	.5306	35	.9734	4	.5571	40	.4429	10
10 0	.2397	71	.9934	3	.2463	73	.7537	80 0	20 0	.5341	34	.9730	5	.5611	39	.4389	70 0
10	.2468	70	.9931	2	.2536	73	.7464	50	10	.5375	34	.9725	4	.5650	39	.4350	50
20	.2538	68	.9929	2	.2609	71	.7391	40	20	.5409	34	.9721	5	.5689	38	.4311	40
30	.2606	68	.9927	3	.2680	70	.7320	30	30	.5443	34	.9716	5	.5727	39	.4273	30
40	.2674	66	.9924	2	.2750	69	.7250	20	40	.5477	33	.9711	5	.5766	38	.4234	20
50	.2740	66	.9922	3	.2819	68	.7181	10	50	.5510	33	.9706	4	.5804	38	.4196	10
11 0	.2806	64	.9919	2	.2887	66	.7113	79 0	21 0	.5543	33	.9702	5	.5842	37	.4158	69 0
10	.2870	64	.9917	3	.2953	67	.7047	50	10	.5576	33	.9697	5	.5879	38	.4121	50
20	.2934	63	.9914	2	.3020	65	.6980	40	20	.5609	32	.9692	5	.5917	37	.4083	40
30	.2997	61	.9912	3	.3085	64	.6915	30	30	.5641	32	.9687	5	.5954	37	.4046	30
40	.3058	61	.9909	2	.3149	63	.6851	20	40	.5673	31	.9682	5	.5991	37	.4009	20
50	.3119	60	.9907	3	.3212	63	.6788	10	50	.5704	32	.9677	5	.6028	36	.3972	10
12 0	.3179	59	.9904	3	.3275	61	.6725	78 0	22 0	.5736	31	.9672	5	.6064	36	.3936	68 0
10	.3238	58	.9901	2	.3336	61	.6664	50	10	.5767	31	.9667	6	.6100	36	.3900	50
20	.3296	57	.9899	3	.3397	61	.6603	40	20	.5798	30	.9661	5	.6136	36	.3864	40
30	.3353	57	.9896	3	.3458	59	.6542	30	30	.5828	31	.9656	5	.6172	36	.3828	30
40	.3410	56	.9893	3	.3517	59	.6483	20	40	.5859	30	.9651	5	.6208	35	.3792	20
50	.3466	53	.9890	3	.3576	58	.6424	10	50	.5889	30	.9646	6	.6243	36	.3757	10
13 0	.3521	54	.9887	3	.3634	57	.6366	77 0	23 0	.5919	29	.9640	5	.6279	35	.3721	67 0
10	.3575	54	.9884	3	.3691	57	.6309	50	10	.5948	30	.9635	6	.6314	34	.3686	50
20	.3629	53	.9881	3	.3748	56	.6252	40	20	.5978	29	.9629	5	.6348	35	.3652	40
30	.3682	52	.9878	3	.3804	55	.6196	30	30	.6007	29	.9624	6	.6383	34	.3617	30
40	.3734	52	.9875	3	.3859	55	.6141	20	40	.6036	29	.9618	5	.6417	35	.3583	20
50	.3786	51	.9872	3	.3914	54	.6086	10	50	.6065	28	.9613	6	.6452	34	.3548	10
14 0	.3837	50	.9869	3	.3968	53	.6032	76 0	24 0	.6093	28	.9607	5	.6486	34	.3514	66 0
10	.3887	50	.9866	3	.4021	53	.5979	50	10	.6121	28	.9602	6	.6520	33	.3480	50
20	.3937	49	.9863	4	.4074	53	.5926	40	20	.6149	28	.9596	6	.6553	34	.3447	40
30	.3986	49	.9859	3	.4127	52	.5873	30	30	.6177	28	.9590	6	.6587	33	.3413	30
40	.4035	48	.9856	3	.4178	51	.5822	20	40	.6205	27	.9584	5	.6620	34	.3380	20
50	.4083	47	.9853	4	.4230	51	.5770	10	50	.6232	27	.9579	6	.6654	33	.3346	10
15 0	9.4130	47	9.9849	3	9.4281	50	0.5719	75 0	25 0	0.6259	27	9.9573	7	9.6687	33	0.3313	65 0
Arc.	Cos.	Df.	Sin.	Df.	Cot.	Df.	Tan.	Arc.	Arc.	Cos.	Df.	Sin.	Df.	Cot.	Df.	Tan.	Arc.

TABLE IIIA—*Continued.*

LOGARITHMS OF SINES AND TANGENTS.

Arc. ° ′	Sin.	Df.	Cos.	Df.	Tan.	Df.	Cot.	Arc. ° ′	Arc. ° ′	Sin.	Df.	Cos.	Df.	Tan.	Df.	Cot.	Arc. ° ′
25 0	9.6259	27	9.9573	6	9.6687	33	0.3313	65 0	35 0	9.7586	18	9.9134	9	9.8452	27	0.1548	55 0
10	.6286	27	.9567	6	.6720	32	.3280	50	10	.7604	18	.9125	9	.8479	27	.1521	50
20	.6313	27	.9561	6	.6752	33	.3248	40	20	.7622	18	.9116	9	.8506	27	.1494	40
30	.6340	26	.9555	6	.6785	32	.3215	30	30	.7640	17	.9107	9	.8533	26	.1467	30
40	.6366	26	.9549	6	.6817	33	.3183	20	40	.7657	18	.9098	9	.8559	27	.1441	20
50	.6392	26	.9543	6	.6850	32	.3150	10	50	.7675	17	.9089	9	.8586	27	.1414	10
26 0	.6418	26	.9537	7	.6882	32	.3118	64 0	36 0	.7692	18	.9080	10	.8613	26	.1387	54 0
10	.6444	26	.9530	6	.6914	32	.3086	50	10	.7710	17	.9070	9	.8639	27	.1361	50
20	.6470	25	.9524	6	.6946	31	.3054	40	20	.7727	17	.9061	9	.8666	26	.1334	40
30	.6495	26	.9518	6	.6977	32	.3023	30	30	.7744	17	.9052	10	.8692	26	.1308	30
40	.6521	25	.9512	7	.7009	31	.2991	20	40	.7761	17	.9042	9	.8718	27	.1282	20
50	.6546	24	.9505	6	.7040	32	.2960	10	50	.7778	17	.9033	10	.8745	26	.1255	10
27 0	.6570	25	.9499	7	.7072	31	.2928	63 0	37 0	.7795	16	.9023	9	.8771	26	.1229	53 0
10	.6595	25	.9492	6	.7103	31	.2897	50	10	.7811	17	.9014	10	.8797	27	.1203	50
20	.6620	24	.9486	7	.7134	31	.2866	40	20	.7828	16	.9004	9	.8824	26	.1176	40
30	.6644	24	.9479	6	.7165	31	.2835	30	30	.7844	17	.8995	10	.8850	26	.1150	30
40	.6668	24	.9473	7	.7196	30	.2804	20	40	.7861	16	.8985	10	.8876	26	.1124	20
50	.6692	24	.9466	7	.7226	31	.2774	10	50	.7877	16	.8975	10	.8902	26	.1098	10
28 0	.6716	24	.9459	6	.7257	30	.2743	62 0	38 0	.7893	17	.8965	10	.8928	26	.1072	52 0
10	.6740	23	.9453	7	.7287	30	.2713	50	10	.7910	16	.8955	10	.8954	26	.1046	50
20	.6763	24	.9446	7	.7317	31	.2683	40	20	.7926	15	.8945	10	.8980	26	.1020	40
30	.6787	23	.9439	7	.7348	30	.2652	30	30	.7941	16	.8935	10	.9006	26	.0994	30
40	.6810	23	.9432	7	.7378	30	.2622	20	40	.7957	16	.8925	10	.9032	26	.0968	20
50	.6833	23	.9425	7	.7408	30	.2592	10	50	.7973	16	.8915	10	.9058	26	.0942	10
29 0	.6856	22	.9418	7	.7438	29	.2562	61 0	39 0	.7989	15	.8905	10	.9084	26	.0916	51 0
10	.6878	23	.9411	7	.7467	30	.2533	50	10	.8004	16	.8895	11	.9110	25	.0890	50
20	.6901	22	.9404	7	.7497	29	.2503	40	20	.8020	15	.8884	10	.9135	26	.0865	40
30	.6923	23	.9397	7	.7526	30	.2474	30	30	.8035	15	.8874	10	.9161	26	.0839	30
40	.6946	22	.9390	7	.7556	29	.2444	20	40	.8050	16	.8864	11	.9187	25	.0813	20
50	.6968	22	.9383	8	.7585	29	.2415	10	50	.8066	15	.8853	10	.9212	26	.0788	10
30 0	.6990	22	.9375	7	.7614	30	.2386	60 0	40 0	.8081	15	.8843	11	.9238	26	.0762	50 0
10	.7012	21	.9368	7	.7644	29	.2356	50	10	.8096	15	.8832	11	.9264	25	.0736	50
20	.7033	22	.9361	8	.7673	28	.2327	40	20	.8111	14	.8821	11	.9289	26	.0711	40
30	.7055	21	.9353	7	.7701	29	.2299	30	30	.8125	15	.8810	10	.9315	26	.0685	30
40	.7076	21	.9346	8	.7730	29	.2270	20	40	.8140	15	.8800	11	.9341	25	.0659	20
50	.7097	21	.9338	7	.7759	29	.2241	10	50	.8155	14	.8789	11	.9366	26	.0634	10
31 0	.7118	21	.9331	8	.7788	28	.2212	59 0	41 0	.8169	15	.8778	11	.9392	25	.0608	49 0
10	.7139	21	.9323	8	.7816	29	.2184	50	10	.8184	14	.8767	11	.9417	26	.0583	50
20	.7160	21	.9315	7	.7845	28	.2155	40	20	.8198	15	.8756	11	.9443	25	.0557	40
30	.7181	20	.9308	8	.7873	29	.2127	30	30	.8213	14	.8745	12	.9468	26	.0532	30
40	.7201	21	.9300	8	.7902	28	.2098	20	40	.8227	14	.8733	11	.9494	25	.0506	20
50	.7222	20	.9292	8	.7930	28	.2070	10	50	.8241	14	.8722	11	.9519	25	.0481	10
32 0	.7242	20	.9284	8	.7958	28	.2042	58 0	42 0	.8255	14	.8711	12	.9544	26	.0456	48 0
10	.7262	20	.9276	8	.7986	28	.2014	50	10	.8269	14	.8699	11	.9570	25	.0430	50
20	.7282	20	.9268	8	.8014	28	.1986	40	20	.8283	14	.8688	12	.9595	26	.0405	40
30	.7302	20	.9260	8	.8042	28	.1958	30	30	.8297	14	.8676	11	.9621	25	.0379	30
40	.7322	20	.9252	8	.8070	27	.1930	20	40	.8311	13	.8665	12	.9646	25	.0354	20
50	.7342	19	.9244	8	.8097	28	.1903	10	50	.8324	14	.8653	12	.9671	26	.0329	10
33 0	.7361	19	.9236	8	.8125	28	.1875	57 0	43 0	.8338	13	.8641	12	.9697	25	.0303	47 0
10	.7380	20	.9228	9	.8153	27	.1847	50	10	.8351	14	.8629	11	.9722	25	.0278	50
20	.7400	19	.9219	8	.8180	28	.1820	40	20	.8365	13	.8618	12	.9747	25	.0253	40
30	.7419	19	.9211	8	.8208	27	.1792	30	30	.8378	13	.8606	12	.9772	26	.0228	30
40	.7438	19	.9203	9	.8235	28	.1765	20	40	.8391	14	.8594	12	.9798	25	.0202	20
50	.7457	19	.9194	8	.8263	27	.1737	10	50	.8405	13	.8582	13	.9823	25	.0177	10
34 0	.7476	18	.9186	9	.8390	27	.1710	56 0	44 0	.8418	13	.8569	12	.9848	26	.0152	46 0
10	.7494	19	.9177	8	.8317	27	.1683	50	10	.8431	13	.8557	12	.9874	25	.0126	50
20	.7513	18	.9169	9	.8344	27	.1656	40	20	.8444	13	.8545	13	.9899	25	.0101	40
30	.7531	19	.9160	9	.8371	27	.1629	30	30	.8457	12	.8532	12	.9924	25	.0076	30
40	.7550	18	.9151	9	.8398	27	.1602	20	40	.8469	13	.8520	13	.9949	26	.0051	20
50	.7568	18	.9142	8	.8425	27	.1575	10	50	.8482	13	.8507	12	9.9975	25	.0025	10
35 0	9.7586	18	9.9134	9	9.8452	27	0.1548	55 0	45 0	9.8495		9.8495		0.0000		0.0000	45 0
Arc.	Cos.	Df.	Sin.	Df.	Cot.	Df.	Tan.	Arc.	Arc.	Cos.	Df.	Sin.	Df.	Cot.	Df.	Tan.	Arc.

TABLE IV.

LOGARITHMIC TRAVERSE TABLE. § 173.

Zero angle at South Point, and increasing to W. (90°), N. (180°), E. (270°).

Arc 1st and 3d. Quadrants.	Log. sin. (Dep.)	Log. cos. (Lat.)	Arc 2d and 4th. Quadrants.	Arc 1st and 3d. Quadrants.	Log. sin. (Dep.)	Log. cos. (Lat.)	Arc 2d and 4th. Quadrants.	Arc 1st and 3d. Quadrants.	Log. sin. (Dep.)	Log. cos. (Lat.)	Arc 2d and 4th. Quadrants.
0° 180°		**10.0000**	**180° 360°**	**1° 181°**	**8.2419**	**9.9990**	**179° 359°**	**2° 182°**	**8.5428**	**9.9997**	**178° 358°**
1′	6.4637	.0000	59′	1′	.2490	.9999	59′	1′	.5464	.9997	59′
2	.7648	.0000	58	2	.2561	.9999	58	2	.5500	.9997	58
3	6.9404	.0000	57	3	.2630	.9999	57	3	.5535	.9997	57
4	7.0658	.0000	56	4	.2699	.9999	56	4	.5571	.9997	56
— 5 —	.1627	.0000	— 55 —	— 5 —	.2766	.9999	— 55 —	— 5 —	.5605	.9997	— 55 —
6	.2419	.0000	54	6	.2832	.9999	54	6	.5640	.9997	54
7	.3088	.0000	53	7	.2898	.9999	53	7	.5674	.9997	53
8	.3668	.0000	52	8	.2962	.9999	52	8	.5708	.9997	52
9	.4180	.0000	51	9	.3025	.9999	51	9	.5742	.9997	51
10	**7.4637**	**10.0000**	**50**	**10**	**8.3088**	**9.9909**	**50**	**10**	**8.5776**	**9.9997**	**50**
11	.5051	.0000	49	11	.3150	.9999	49	11	.5809	.9997	49
12	.5429	.0000	48	12	.3210	.9999	48	12	.5842	.9997	48
13	.5777	0000	47	13	.3270	.9999	47	13	.5875	.9997	47
14	.6099	.0000	46	14	.3229	.9999	46	14	.5907	.9997	46
— 15 —	.6398	.0000	— 45 —	— 15 —	.3388	.9999	— 45 —	— 15 —	.5939	.9997	— 45 —
16	.6678	.0000	44	16	.3445	.9999	44	16	.5972	.9997	44
17	.6942	.0000	43	17	.3502	.9999	43	17	.6003	.9997	43
18	.7190	.0000	[illegible]	18	.3558	.9999	42	18	.6035	.9996	42
19	.7425	.0000	4[illegible]	19	.3613	.9999	41	19	.6066	.9996	41
20	**7.7648**	**10.0000**	**4[illegible]**	**20**	**8.3668**	**9.9999**	**40**	**20**	**8.6097**	**9.9996**	**40**
21	.7859	.0000	39	21	.3722	.9999	39	21	.6128	.9996	39
22	.8061	.0000	38	22	.3775	.9999	38	22	.6159	.9996	38
23	.8255	.0000	37	23	.3828	.9999	37	23	.6189	.9996	37
24	.8439	.0000	36	24	.3880	.9999	36	24	.6220	.9996	36
— 25 —	.8617	.0000	— 35 —	— 25 —	.3931	.9999	— 35 —	— 25 —	.6250	.9996	— 35 —

26	.8787	.0000	34	26	.3982	.9999	34	26	.6279	.9996	34
27	.8951	.0000	33	27	.4032	.9999	33	27	.6309	.9996	33
28	.9109	.0000	32	28	.4082	.9999	32	28	.6339	.9996	32
29	.9261	.0000	31	29	.4131	.9999	31	29	.6368	.9996	31
30	**7.9408**	**10.0000**	**30**	**30**	**8.4179**	**9.9999**	**30**	**30**	**8.6397**	**9.9996**	**30**
31	.9551	.0000	29	31	.4227	.9998	29	31	.6426	.9996	29
32	.9689	.0000	28	32	.4275	.9998	28	32	.6454	.9996	28
33	.9822	.0000	27	33	.4322	.9998	27	33	.6483	.9996	27
34	7.9952	.0000	26	34	.4368	.9998	26	34	.6511	.9996	26
— 35 —	8.0078	.0000	— 25 —	— 35 —	.4414	.9998	— 25 —	— 35 —	.6539	.9966	— 25 —
36	.0200	.0000	24	36	.4459	.9998	24	36	.6567	.9996	24
37	.0319	.0000	23	37	.4504	.9998	23	37	.6595	.9995	23
38	.0435	.0000	22	38	.4549	.9998	22	38	.6622	.9995	22
39	.0548	.0000	21	39	.4593	.9998	21	39	.6650	.9995	21
40	**8.0658**	**10.0000**	**20**	**40**	**8.4637**	**9.9998**	**20**	**40**	**8.6677**	**9.9995**	**20**
41	.0765	.0000	19	41	.4680	.9998	19	41	.6704	.9995	19
42	.0870	.0000	18	42	.4723	.9998	18	42	.6731	.9995	18
43	.0972	.0000	17	43	.4765	.999[illegible]	17	43	.6758	.9995	17
44	.1072	.0000	16	44	.4807	.9998	16	44	.6784	.9995	16
— 45 —	.1169	.0000	— 15 —	— 45 —	.4848	.9998	— 15 —	— 45 —	.6810	.9995	— 15 —
46	.1265	.0000	14	46	.4890	.9998	14	46	.6837	.9995	14
47	.1358	.0000	13	47	.4930	.9998	13	47	.6863	.9995	13
48	.1450	.0000	12	48	.4971	.9998	12	48	.6889	.9995	12
49	.1539	.0000	11	49	.5011	.9998	11	49	.6914	.9995	11
50	**8.1627**	**10.0000**	**10**	**50**	**8.5050**	**9.9998**	**10**	**50**	**8.6940**	**9.9995**	**10**
51	.1713	.0000	9	51	.5090	.9998	9	51	.6965	.9995	9
52	.1797	0.0000	8	52	.5129	.9998	8	52	.6991	.9995	8
53	.1880	9.9999	7	53	.5167	.9998	7	53	.7016	.9994	7
54	.1961	.9999	6	54	.5206	.9998	6	54	.7041	.9994	6
— 55 —	.2041	.9999	— 5 —	— 55 —	.5243	.9998	— 5 —	— 55 —	.7066	.9994	— 5 —
56	.2119	.9999	4	56	.5281	.9998	4	56	.7090	.9994	4
57	.2196	.9999	3	57	.5318	.9997	3	57	.7115	.9994	3
58	.2271	.9999	2	58	.5355	.9997	2	58	.7140	.9994	2
59	.2346	.9999	1	59	.5392	.9997	1	59	.7164	.9994	1
1° 181°	**8.2419**	**9.9999**	**179° 359°**	**2° 182°**	**8.5428**	**9.9997**	**178° 358°**	**3° 183°**	**8.7188**	**9.9994**	**177° 357°**

TABLE IV.—*Continued.*

Logarithmic Traverse Table.

Zero angle at South Point, and increasing to W. (90°), N. (180°), E. (270°).

Arc 1st and 3d. Quad-rants.	Log. sin. (Dep.)	Sin. Dif. for 1′.	Log. cos. (Lat.)	Arc 2d and 4th. Quadrants.	Arc 1st and 3d. Quad-rants.	Log. sin. (Dep.)	Sin. Dif. for 1′.	Log. cos. (Lat.)	Arc 2d and 4th. Quadrants.	Arc 1st and 3d. Quad-rants.	Log. sin. (Dep)	Sin. Dif. for 1′.	Log. cos. (Lat.)	Arc 2d and 4th. Quadrants.
3° 183°	8.7188	23.5	9.9994	**177° 357°**	**17° 197°**	9.4659	4.1	9.9806	**163° 343°**	**31° 211°**	9.7118	2.1	9.9331	**149° 329°**
10′	.7423	22.2	.9993	50′	10′	.4700	4.1	.9802	50′	10′	.7139	2.1	.9323	50′
20	.7645	21.2	.9993	40	20	.4741	4.0	.9798	40	20	.7160	2.1	.9315	40
— 30 —	.7857	20.2	.9992	— 30 —	— 30 —	.4781	4.0	.9794	— 30 —	— 30 —	.7181	2.0	.9308	— 30 —
40	.8059	19.2	.9991	20	40	.4821	4.0	.9790	20	40	.7201	2.1	.9300	20
50	.8251	18.5	.9990	10	50	.4861	3.9	.9786	10	50	.7222	2.0	.9292	10
4° 184°	8.8436	17.7	9.9989	**176° 356°**	**18° 198°**	9.4900	3.9	9.9782	**162° 342°**	**32° 212°**	9.7242	2.0	9.9284	**148° 328°**
10	.8613	17.0	.9989	50	10	.4939	3.8	.9778	50	10	.7262	2.0	.9276	50
20	.8783	16.3	.9988	40	20	.4977	3.8	.9774	40	20	.7282	2.0	.9268	40
— 30 —	.8946	15.8	.9987	— 30 —	— 30 —	.5015	3.7	.9770	— 30 —	— 30 —	.7302	2.0	.9260	— 30 —
40	.9104	15.2	.9986	20	40	.5052	3.8	.9765	20	40	.7322	2.0	.9252	20
50	.9256	14.7	.9985	10	50	.5090	3.6	.9761	10	50	.7342	1.9	.9244	10
5° 185°	8.9403	14.2	9.9983	**175° 355°**	**19° 199°**	9.5126	3.7	9.9757	**161° 341°**	**33° 213°**	9.7361	1.9	9.9236	**147° 327°**
10	.9545	13.7	.9982	50	10	.5163	3.6	.9752	50	10	.7380	2.0	.9228	50
20	.9682	13.4	.9981	40	20	.5199	3.6	.9748	40	20	.7400	1.9	.9219	40
— 30 —	.9816	13.1	.9980	— 30 —	— 30 —	.5235	3.5	.9743	— 30 —	— 30 —	.7419	1.9	.9211	— 30 —
40	8.9945	12.5	.9979	20	40	.5270	3.6	.9739	20	40	.7438	1.9	.9203	20
50	9.0070	12.2	.9977	10	50	.5306	3.5	.9734	10	50	.7457	1.9	.9194	10
6° 186°	9.0192	11.9	9.9975	**174° 354°**	**20° 200°**	9.5341	3.4	9.9730	**160° 340°**	**34° 214°**	9.7476	1.8	9.9186	**146° 326°**
10	.0311	11.5	.9975	50	10	.5375	3.4	.9725	50	10	.7494	1.9	.9177	50
20	.0426	11.3	.9973	40	20	.5409	3.4	.9721	40	20	.7513	1.8	.9169	40
— 30 —	.0539	10.9	.9972	— 30 —	— 30 —	.5443	3.4	.9716	— 30 —	— 30 —	.7531	1.9	.9160	— 30 —
40	.0648	10.7	.9971	20	40	.5477	3.3	.9711	20	40	.7550	1.8	.9151	20
50	.0755	10.4	.9969	10	50	.5510	3.3	.9706	10	50	.7568	1.8	.9142	10
7° 187°	9.0859	10.2	9.9968	**173° 353°**	**21° 201°**	9.5543	3.3	9.9702	**159° 339°**	**35° 215°**	9.7586	1.8	9.9134	**145° 325°**
10	.0961	9.9	.9966	50	10	.5576	3.3	.9697	50	10	.7604	1.8	.9125	50
20	.1060	9.7	.9964	40	20	.5609	3.2	.9692	40	20	.7622	1.8	.9116	40
— 30 —	.1157	9.5	.9963	— 30 —	— 30 —	.5641	3.2	.9687	— 30 —	— 30 —	.7640	1.7	.9107	— 30 —
40	.1252	9.3	.9961	20	40	.5673	3.1	.9682	20	40	.7657	1.8	.9098	20
50	.1345	9.1	.9959	10	50	.5704	3.2	.9677	10	50	.7675	1.7	.9089	10
8° 188°	9.1436	8.9	9.9958	**172° 352°**	**22° 202°**	9.5736	3.1	9.9672	**158° 338°**	**36° 216°**	9.7692	1.8	9.9080	**144° 324°**
10	.1525	8.7	.9956	50	10	.5767	3.1	.9667	50	10	.77·0	1.7	.9070	50
20	.1612	8.5	.9954	40	20	.5798	3.0	.9661	40	20	.7727	1.7	.9061	40
— 30 —	.1697	8.4	.9952	— 30 —	— 30 —	.5828	3.1	.9656	— 30 —	— 30 —	.7744	1.7	.9052	— 30 —
40	.1781	8.2	.9950	20	40	.5859	3.0	.9651	20	40	.7761	1.7	.9042	20
50	.1863	8.0	.9948	10	50	.5889	3.0	.9646	10	50	.7778	1.7	.9033	10
9° 189°	9.1943		9.9946	**171° 351°**	**23° 203°**	9.5919		9.9640	**157° 337°**	**37° 217°**	9.7795		9.9023	**143° 323°**

9° 189°	9.1943	7.9	9.9946	**171° 351°**	**23° 203°**	9.5919	2.9	9.9640	**157° 337°**	**37° 217°**	9.7795	1.6	9.9023	**143° 323°**
10	.2022	7.8	.9944	50	10	.5948	3.0	.9635	50	10	.7811	1.7	.9014	50
20	.2100	7.6	.9942	40	20	.5978	2.9	.9629	40	20	.7828	1.6	.9004	40
— 30 —	.2176	7.5	.9940	— 30 —	— 30 —	.6007	2.9	.9624	— 30 —	— 30 —	.7844	1.7	.8995	— 30 —
40	.2251	7.3	.9938	20	40	.6036	2.9	.9618	20	40	.7861	1.6	.8985	20
50	.2324	7.3	.9936	10	50	.6065	2.8	.9613	10	50	.7877	1.6	.8975	10
10° 190°	9.2397	7.1	9.9934	**170° 350°**	**24° 204°**	9.6093	2.8	9.9607	**156° 336°**	**38° 218°**	9.7893	1.7	9.8965	**142° 322°**
10	.2468	7.0	.9931	50	10	.6121	2.8	.9602	50	10	.7910	1.6	.8955	50
20	.2538	6.8	.9929	40	20	.6149	2.8	.9596	40	20	.7926	1.5	.8945	40
— 30 —	.2606	6.8	.9927	— 30 —	— 30 —	.6177	2.8	.9590	— 30 —	— 30 —	.7941	1.6	.8935	— 30 —
40	.2674	6.6	.9924	20	40	.6205	2.7	.9584	20	40	.7957	1.6	.8925	20
50	.2740	6.6	.9922	10	50	.6232	2.7	.9579	10	50	.7973	1.6	.8915	10
11° 191°	9.2806	6.4	9.9919	**169° 349°**	**25° 205°**	9.6259	2.7	9.9573	**155° 335°**	**39° 219°**	9.7989	1.5	9.8905	**141° 321°**
10	.2870	6.4	.9917	50	10	.6286	2.7	.9567	50	10	.8004	1.6	.8895	50
20	.2934	6.3	.9914	40	20	.6313	2.7	.9561	40	20	.8020	1.5	.8884	40
— 30 —	.2997	6.1	.9912	— 30 —	— 30 —	.6340	2.6	.9555	— 30 —	— 30 —	.8035	1.5	.8874	— 30 —
40	.3058	6.1	.9909	20	40	.6366	2.6	.9549	20	40	.8050	1.6	.8864	20
50	.3119	6.0	.9907	10	50	.6392	2.6	.9543	10	50	.8066	1.5	.8853	10
12° 192°	9.3179	5.9	9.9904	**168° 348°**	**26° 206°**	9.6418	2.6	9.9537	**154° 334°**	**40° 220°**	9.8081	1.5	9.8843	**140° 320°**
10	.3238	5.8	.9901	50	10	.6444	2.6	.9530	50	10	.8096	1.5	.8832	50
20	.3296	5.7	.9899	40	20	.6470	2.5	.9524	40	20	.8111	1.4	.8821	40
— 30 —	.3353	5.7	.9896	— 30 —	— 30 —	.6495	2.6	.9518	— 30 —	— 30 —	.8125	1.5	.8810	— 30 —
40	.3410	5.6	.9893	20	40	.6521	2.5	.9512	20	40	.8140	1.5	.8800	20
50	.3466	5.5	.9890	10	50	.6546	2.4	.9505	10	50	.8155	1.4	.8789	10
13° 193°	9.3521	5.4	9.9887	**167° 347°**	**27° 207°**	9.6570	2.5	9.9499	**153° 333°**	**41° 221°**	9.8169	1.5	9.8778	**139° 319°**
10	.3575	5.4	.9884	50	10	.6595	2.5	.9492	50	10	.8184	1.4	.8767	50
20	.3629	5.3	.9881	40	20	.6620	2.4	.9486	40	20	.8198	1.5	.8756	40
— 30 —	.3682	5.2	.9878	— 30 —	— 30 —	.6644	2.4	.9479	— 30 —	— 30 —	.8213	1.4	.8745	— 30 —
40	.3734	5.2	.9875	20	40	.6668	2.4	.9473	20	40	.8227	1.4	.8733	20
50	.3786	5.1	.9872	10	50	.6692	2.4	.9466	10	50	.8241	1.4	.8722	10
14° 194°	9.3837	5.0	9.9869	**166° 346°**	**28° 208°**	9.6716	2.4	9.9459	**152° 332°**	**42° 222°**	9.8255	1.4	9.8711	**138° 318°**
10	.3887	5.0	.9866	50	10	.6740	2.3	.9453	50	10	.8269	1.4	.8699	50
20	.3937	4.9	.9863	40	20	.6763	2.4	.9446	40	20	.8283	1.4	.8688	40
— 30 —	.3986	4.9	.9859	— 30 —	— 30 —	.6787	2.3	.9439	— 30 —	— 30 —	.8297	1.4	.8676	— 30 —
40	.4035	4.8	.9856	20	40	.6810	2.3	.9432	20	40	.8311	1.3	.8665	20
50	.4083	4.7	.9853	10	50	.6833	2.3	.9425	10	50	.8324	1.4	.8653	10
15° 195°	9.4130	4.7	9.9849	**165° 345°**	**29° 209°**	9.6856	2.2	9.9418	**151° 331°**	**43° 223°**	9.8338	1.3	9.8641	**137° 317°**
10	.4177	4.6	.9846	50	10	.6878	2.3	.9411	50	10	.8351	1.4	.8629	50
20	.4223	4.6	.9843	40	20	.6901	2.2	.9404	40	20	.8365	1.3	.8618	40
— 30 —	.4269	4.5	.9839	— 30 —	— 30 —	.6923	2.3	.9397	— 30 —	— 30 —	.8378	1.3	.8606	— 30 —
40	.4314	4.5	.9836	20	40	.6946	2.2	.9390	20	40	.8391	1.4	.8594	20
50	.4359	4.4	.9832	10	50	.6968	2.2	.9383	10	50	.8405	1.3	.8582	10
16° 196°	9.4403	4.4	9.9828	**164° 344°**	**30° 210°**	9.6990	2.2	9.9375	**150° 330°**	**44° 224°**	9.8418	1.3	9.8569	**136° 316°**
10	.4447	4.4	.9825	50	10	.7012	2.1	.9368	50	10	.8431	1.3	.8557	50
20	.4491	4.2	.9821	40	20	.7033	2.2	.9361	40	20	.8444	1.3	.8545	40
— 30 —	.4533	4.3	.9817	— 30 —	— 30 —	.7055	2.1	.9353	— 30 —	— 30 —	.8457	1.2	.8532	— 30 —
40	.4576	4.2	.9814	20	40	.7076	2.1	.9346	20	40	.8469	1.3	.8520	20
50	.4618	4.1	.9810	10	50	.7097	2.1	.9338	10	50	.8482	1.3	.8507	10
17° 197°	9.4659		9.9806	**163° 343°**	**31° 211°**	9.7118		9.9331	**149° 329°**	**45° 225°**	9.8495		9.8495	**135° 315°**

TABLE IV.—*Continued.*

Logarithmic Traverse Table.

Zero angle at South Point, and increasing to W. (90°), N. (180°), E. (270°).

Arc 1st and 3d Quadrants.	Log. sin. (Dep.)	Cos. Dif. for 1'.	Log. cos. (Lat.)	Arc 2d and 4th Quadrants.	Arc 1st and 3d Quadrants.	Log. sin. (Dep.)	Cos. Dif. for 1'.	Log. cos. (Lat.)	Arc 2d and 4th Quadrants.	Arc 1st and 3d Quadrants.	Log. sin. (Dep.)	Cos. Dif. for 1'.	Log. cos. (Lat.)	Arc 2d and 4th Quadrants.
45° 225°	9.8495		9.8495	**135° 315°**	**59° 239°**	9.9331		9.7118	**121° 301°**	**73° 253°**	9.9806		9.4659	**107° 287°**
10'	.8507	1.3	.8482	50'	10'	.9338	2.1	.7097	50'	10'	.9810	4.1	.4618	50'
20	.8520	1.3	.8469	40	20	.9346	2.1	.7076	40	20	.9814	4.2	.4576	40
— 30 —	.8532	1.2	.8457	— 30 —	— 30 —	.9353	2.1	.7055	— 30 —	— 30 —	.9817	4.3	.4533	— 30 —
40	.8545	1.3	.8444	20	40	.9361	2.2	.7033	20	40	.9821	4.2	.4491	20
50	.8557	1.3	.8431	10	50	.9368	2.1	.7012	10	50	.9825	4.4	.4447	10
46° 226°	9.8569	1.3	9.8418	**134° 314°**	**60° 240°**	9.9375	2.2	9.6990	**120° 300°**	**74° 254°**	9.9829	4.4	9.4403	**106° 286°**
10	.8582	1.3	.8405	50	10	.9383	2.2	.6968	50	10	.9832	4.4	.4359	50
20	.8594	1.4	.8391	40	20	.9390	2.2	.6946	40	20	.9836	4.5	.4314	40
— 30 —	.8606	1.3	.8378	— 30 —	— 30 —	.9397	2.3	.6923	— 30 —	— 30 —	.9839	4.5	.4269	— 30 —
40	.8618	1.3	.8365	20	40	.9404	2.2	.6901	20	40	.9843	4.6	.4223	20
50	.8629	1.4	.8351	10	50	.9411	2.3	.6878	10	50	.9846	4.6	.4177	10
47° 227°	9.8641	1.3	9.8338	**133° 313°**	**61° 241°**	9.9418	2.2	9.6856	**119° 299°**	**75° 255°**	9.9849	4.7	9.4130	**105° 285°**
10	.8653	1.4	.8324	50	10	.9425	2.3	.6833	50	10	.9853	4.7	.4083	50
20	.8665	1.3	.8311	40	20	.9432	2.3	.6810	40	20	.9856	4.8	.4035	40
— 30 —	.8676	1.4	.8297	— 30 —	— 30 —	.9439	2.3	.6787	— 30 —	— 30 —	.9859	4.9	.3986	— 30 —
40	.8688	1.4	.8283	20	40	.9446	2.4	.6763	20	40	.9863	4.9	.3937	20
50	.8699	1.4	.8269	10	50	.9453	2.3	.6740	10	50	.9866	5.0	.3887	10
48° 228°	9.8711	1.4	9.8255	**132° 312°**	**62° 242°**	9.9459	2.4	9.6716	**118° 298°**	**76° 256°**	9.9869	5.0	9.3857	**104° 284°**
10	.8722	1.4	.8241	50	10	.9466	2.4	.6692	50	10	.9872	5.1	.3786	50
20	.8733	1.4	.8227	40	20	.9473	2.4	.6668	40	20	.9875	5.2	.3734	40
— 30 —	.8745	1.4	.8213	— 30 —	— 30 —	.9479	2.4	.6644	— 30 —	— 30 —	.9878	5.2	.3682	— 30 —
40	.8756	1.5	.8198	20	40	.9486	2.4	.6620	20	40	.9881	5.3	.3629	20
50	.8767	1.4	.8184	10	50	.9492	2.5	.6595	10	50	.9884	5.4	.3575	10
49° 229°	9.8778	1.5	9.8169	**131° 311°**	**63° 243°**	9.9499	2.5	9.6570	**117° 297°**	**77° 257°**	9.9887	5.4	9.3521	**103° 283°**
10	.8789	1.4	.8155	50	10	.9505	2.4	.6546	50	10	.9890	5.5	.3466	50
20	.8800	1.5	.8140	40	20	.9512	2.5	.6521	40	20	.9843	5.6	.3410	40
— 30 —	.8810	1.5	.8125	— 30 —	— 30 —	.9518	2.6	.6495	— 30 —	— 30 —	.9896	5.7	.3353	— 30 —
40	.8821	1.4	.8111	20	40	.9524	2.5	.6470	20	40	.9899	5.7	.3296	20
50	.8832	1.5	.8096	10	50	.9530	2.6	.6444	10	50	.9901	5.8	.3238	10
50° 230°	9.8843	1.5	9.8081	**130° 310°**	**64° 244°**	9.9537	2.6	9.6418	**116° 296°**	**78° 258°**	9.9904	5.9	9.3179	**102° 282°**
10	.8853	1.5	.8066	50	10	.9543	2.6	.6392	50	10	.9907	6.0	.3119	50
20	.8864	1.6	.8050	40	20	.9549	2.6	.6366	40	20	.9909	6.1	.3058	40
— 30 —	.8874	1.5	.8035	— 30 —	— 30 —	.9555	2.6	.6340	— 30 —	— 30 —	.9912	6.1	.2997	— 30 —
40	.8884	1.5	.8020	20	40	.9561	2.7	.6313	20	40	.9914	6.3	.2934	20
50	.8895	1.6	.8004	10	50	.9567	2.7	.6286	10	50	.9917	6.4	.2870	10
51° 231°	9.8905	1.5	9.7989	**129° 309°**	**65° 245°**	9.9573	2.7	9.6259	**115° 295°**	**79° 259°**	9.9919	6.4	9.2806	**101° 281°**

51° 231°	9.8905	1.6	9.7989	129° 309°	65° 245°	9.9573	2.7	9.6259	115° 295°	79° 259°	9.9919	6.6	9.2806	101° 281°
10	.8915	1.6	.7973	50	10	.9579	2.7	.6232	50	10	.9922	6.6	.2740	50
20	.8925	1.6	.7957	40	20	.9584	2.8	.6205	40	20	.9924	6.8	.2674	40
— 30 —	.8935	1.5	.7941	— 30 —	— 30 —	.9590	2.8	.6177	— 30 —	— 30 —	.9927	6.8	.2606	— 30 —
40	.8945	1.6	.7925	20	40	.9596	2.8	.6149	20	40	.9929	7.0	.2538	20
50	.8955	1.7	.7910	10	50	.9602	2.8	.6121	10	50	.9931	7.1	.2468	10
52° 232°	9.8965	1.6	9.7893	128° 308°	66° 246°	9.9607	2.8	9.6093	114° 294°	80° 260°	9.9934	7.3	9.2397	100° 280°
10	.8975	1.6	.7877	50	10	.9613	2.9	.6065	50	10	.9936	7.3	.2324	50
20	.8985	1.7	.7861	40	20	.9618	2.9	.6036	40	20	.9938	7.5	.2251	40
— 30 —	.8995	1.6	.7844	— 30 —	— 30 —	.9624	2.9	.6007	— 30 —	— 30 —	.9940	7.6	.2176	— 30 —
40	.9004	1.7	.7823	20	40	.9629	3.0	.5978	20	40	.9942	7.8	.2100	20
50	.9014	1.6	.7811	10	50	.9635	2.9	.5948	10	50	.9944	7.9	.2022	10
53° 233°	9.9023	1.7	9.7795	127° 307°	67° 247°	9.9640	3.0	9.5919	113° 293°	81° 261°	9.9946	8.0	9.1943	99° 279°
10	.9033	1.7	.7778	50	10	.9646	3.0	.5889	50	10	.9948	8.2	.1863	50
20	.9042	1.7	.7761	40	20	.9651	3.1	.5859	40	20	.9950	8.4	.1781	40
— 30 —	.9052	1.7	.7744	— 30 —	— 30 —	.9656	3.0	.5828	— 30 —	— 30 —	.9952	8.5	.1697	— 30 —
40	.9065	1.7	.7727	20	40	.9661	3.1	.5798	20	40	.9954	8.7	.1612	20
50	.9070	1.8	.7710	10	50	.9667	3.1	.5767	10	50	.9956	8.9	.1525	10
54° 234°	9.9080	1.7	9.7692	126° 306°	68° 248°	9.9672	3.2	9.5736	112° 292°	82° 262°	9.9958	9.1	9.1436	98° 278°
10	.9089	1.8	.7675	50	10	.9677	3.1	.5704	50	10	.9959	9.3	.1345	50
20	.9098	1.7	.7657	40	20	.9682	3.2	.5673	40	20	.9961	9.5	.1252	40
— 30 —	.9107	1.8	.7640	— 30 —	— 30 —	.9687	3.2	.5641	— 30 —	— 30 —	.9963	9.7	.1157	— 30 —
40	.9116	1.8	.7622	20	40	.9692	3.3	.5609	20	40	.9964	9.9	.1060	20
50	.9125	1.8	.7604	10	50	.9697	3.3	.5576	10	50	.9966	10.2	.0961	10
55° 235°	9.9134	1.8	9.7586	125° 305°	69° 249°	9.9702	3.3	9.5543	111° 291°	83° 263°	9.9968	10.4	9.0859	97° 277°
10	.9142	1.8	.7568	50	10	.9706	3.3	.5510	50	10	.9969	10.7	.0755	50
20	.9151	1.9	.7550	40	20	.9711	3.4	.5477	40	20	.9971	10.9	.0648	40
— 30 —	.9160	1.8	.7531	— 30 —	— 30 —	.9716	3.4	.5443	— 30 —	— 30 —	.9972	11.3	.0539	— 30 —
40	.9169	1.9	.7513	20	40	.9721	3.4	.5409	20	40	.9973	11.5	.0426	20
50	.9177	1.8	.7494	10	50	.9725	3.4	.5375	10	50	.9975	11.9	.0311	10
56° 236°	9.9186	1.9	9.7476	124° 304°	70° 250°	9.9730	3.5	9.5341	110° 290°	84° 264°	9.9976	12.2	9.0192	96° 276°
10	.9164	1.9	.7457	50	10	.9734	3.6	.5306	50	10	.9977	12.5	9.0070	50
20	.9203	1.9	.7438	40	20	.9739	3.6	.5270	40	20	.9979	12.9	8.9945	40
— 30 —	.9211	1.9	.7419	— 30 —	— 30 —	.9743	3.6	.5235	— 30 —	— 30 —	.9980	13.4	.9816	— 30
40	.9219	2.0	.7400	20	40	.9748	3.6	.5199	20	40	.9981	13.7	.9682	20
50	.9228	1.9	.7380	10	50	.9752	3.5	.5163	10	50	.9982	14.2	.9545	10
57° 237°	9.9236	1.9	9.7361	123° 303°	71° 251°	9.9757	3.6	9.5126	109° 289°	85° 265°	9.9983	14.7	8.9403	95° 275°
10	.9244	2.0	.7342	50	10	.9761	3.8	.5090	50	10	.9985	15.2	.9256	50
20	.9252	2.0	.7322	40	20	.9765	3.7	.5052	40	20	.9986	15.8	.9104	40
— 30 —	.9260	2.0	.7302	— 30 —	— 30 —	.9770	3.8	.5015	— 30 —	— 30 —	.9987	16.3	.8946	— 30 —
40	.9268	2.0	.7282	20	40	.9774	3.8	.4977	20	40	.9988	17.0	.8783	20
50	.9276	2.0	.7262	10	50	.9778	3.9	.4939	10	50	.9989	17.7	.8613	10
58° 238°	9.9284	2.0	9.7242	122° 302°	72° 252°	9.9782	3.9	9.4900	108° 288°	86° 266°	9.9989	18.5	8.8436	94° 274°
10	.9292	2.1	.7222	50	10	.9786	4.0	.4861	50	10	.9990	19.2	.8251	50
20	.9300	2.0	.7201	40	20	.9790	4.0	.4821	40	20	.9991	20.2	.8059	40
— 30 —	.9308	2.1	.7181	— 30 —	— 30 —	.9794	4.0	.4781	— 30 —	— 30 —	.9992	21.2	.7857	— 30 —
40	.9315	2.1	.7160	20	40	.9798	4.1	.4741	20	40	.9993	22.2	.7645	20
50	.9323	2.1	.7139	10	50	.9802	4.1	.4700	10	50	.9993	23.5	.7423	10
59° 239°	9.9331		9.7113	121° 301°	73° 253°	9.9806		9.4659	107° 287°	87° 267°	9.9994		8.7188	93° 273°

TABLE IV.—*Continued.*

LOGARITHMIC TRAVERSE TABLE.

Zero angle at South Point, and increasing to W. (90°), N. (180°), E. (270°).

Arc 1st and 3d Quadrants.	Log. sin. (Dep.)	Log. cos. (Lat.)	Arc 2d and 4th Quadrants.	Arc 1st and 3d Quadrants.	Log. sin. (Dep.)	Log. cos. (Lat.)	Arc 2d and 4th Quadrants.	Arc 1st and 3d Quadrants.	Log. sin. (Dep.)	Log. cos. (Lat.)	Arc 2d and 4th Quadrants.
87° 267°	**9.9994**	**8.7188**	**93° 273°**	**88° 268°**	**9.9997**	**8.5428**	**92° 272°**	**89° 269°**	**9.9999**	**8.2419**	**91° 271°**
1′	.9994	.7164	59′	1′	.9997	.5392	59′	1′	.9999	.2346	59′
2	.9994	.7140	58	2	.9997	.5355	58	2	.9999	.2271	58
3	.9994	.7115	57	3	.9997	.5318	57	3	.9999	.2196	57
4	.9994	.7090	56	4	.9998	.5281	56	4	.9999	.2119	56
5	.9994	.7066	55	5	.9998	.5243	55	5	.9999	.2041	55
6	.9994	.7041	54	6	.9998	.5206	54	6	.9999	.1961	54
7	.9994	.7016	53	7	.9998	.5167	53	7	9.9999	.1880	53
8	.9995	.6991	52	8	.9998	.5129	52	8	10.0000	.1797	52
9	.9995	.6965	51	9	.9998	.5090	51	9	.0000	.1713	51
10	**9.9995**	**8.6940**	**50**	**10**	**9.9998**	**8.5050**	**50**	**10**	**10.0000**	**8.1627**	**50**
11	.9995	.6914	49	11	.9998	.5011	49	11	.0000	.1539	49
12	.9995	.6889	48	12	.9998	.4971	48	12	.0000	.1450	48
13	.9995	.6863	47	13	.9998	.4930	47	13	.0000	.1358	47
14	.9995	.6837	46	14	.9998	.4890	46	14	.0000	.1265	46
15	.9995	.6810	45	15	.9998	.4848	45	15	.0000	.1169	45
16	.9995	.6784	44	16	.9998	.4807	44	16	.0000	.1072	44
17	.9995	.6758	43	17	.9998	.4765	43	17	.0000	.0972	43
18	.9995	.6731	42	18	.9998	.4723	42	18	.0000	.0870	42
19	.9995	.6704	41	19	.9998	.4680	41	19	.0000	.0765	41
20	**9.9995**	**8.6677**	**40**	**20**	**9.9998**	**8.4637**	**40**	**20**	**10.0000**	**8.0658**	**40**
21	.9995	.6650	39	21	.9998	.4593	39	21	.0000	.0548	39
22	.9995	.6622	38	22	.9998	.4549	38	22	.0000	.0435	38
23	.9995	.6595	37	23	.9998	.4504	37	23	.0000	.0319	37
24	.9996	.6567	36	24	.9998	.4459	36	24	.0000	.0200	36
25	.9996	.6539	35	25	.9998	.4418	35	25	.0000	8.0078	35

26	.9996	.6511	34	26	.9998	.4368	34	26	.0000	7.9952	34
27	.9996	.6483	33	27	.9998	.4322	33	27	.0000	.9822	33
28	.9996	.6454	32	28	.9998	.4275	32	28	.0000	.9689	32
29	.9996	.6426	31	29	.9998	.4227	31	29	.0000	.9551	31
30	**9.9996**	**8.6397**	**30**	**30**	**9.9999**	**8.4179**	**30**	**30**	**10.0000**	**7.9408**	**30**
31	.9996	.6368	29	31	.9999	.4131	29	31	.0000	.9261	29
32	.9996	.6339	28	32	.9999	.4082	28	32	.0000	.9109	28
33	.9996	.6309	27	33	.9999	.4032	27	33	.0000	.8951	27
34	.9996	.6279	26	34	.9999	.3982	26	34	.0000	.8787	26
— 35 —	.9996	.6250	— 25 —	— 35 —	.9999	.3931	— 25 —	— 35 —	.0000	.8617	— 25 —
36	.9996	.6220	24	36	.9999	.3880	24	36	.0000	.8439	24
37	.9996	.6189	23	37	.9999	.3828	23	37	.0000	.8255	23
38	.9996	.6159	22	38	.9999	.3775	22	38	.0000	.8061	22
39	.9996	.6128	21	39	.9999	.3722	21	39	.0000	.7859	21
40	**9.9996**	**8.6097**	**20**	**40**	**9.9999**	**8.3668**	**20**	**40**	**10.0000**	**7.7648**	**20**
41	.9996	.6066	19	41	.9999	.3613	19	41	.0000	.7425	19
42	.9996	.6035	18	42	.9999	.3558	18	42	.0000	.7190	18
43	.9997	.6003	17	43	.9999	.3502	17	43	.0000	.6942	17
44	.9997	.5972	16	44	.9999	.3445	16	44	.0000	.6678	16
— 45 —	.9997	.5939	— 15 —	— 45 —	.9999	.3388	— 15 —	— 45 —	.0000	.6398	— 15 —
46	.9997	.5907	14	46	.9999	.3329	14	46	.0000	.6099	14
47	.9997	.5875	13	47	.9999	.3270	13	47	.0000	.5777	13
48	.9997	.5842	12	48	.9999	.3210	12	48	.0000	.5429	12
49	.9997	.5809	11	49	.9999	.3150	11	49	.0000	.5051	11
50	**9.9997**	**8.5776**	**10**	**50**	**9.9999**	**8.3088**	**10**	**50**	**10.0000**	**7.4637**	**10**
51	.9997	.5742	9	51	.9999	.3025	9	51	.0000	.4180	9
52	.9997	.5708	8	52	.9999	.2962	8	52	.0000	.3668	8
53	.9997	.5674	7	53	.9999	.2898	7	53	.0000	.3088	7
54	.0907	.5640	6	54	.9999	.2832	6	54	.0000	.2419	6
— 55 —	.9997	.5605	— 5 —	— 55 —	.9999	.2766	— 5 —	— 55 —	.0000	.1627	— 5 —
56	.9997	.5571	4	56	.9999	.2699	4	56	.0000	7.0658	4
57	.9997	.5535	3	57	.9999	.2630	3	57	.0000	6.9408	3
58	.9997	.5500	2	58	.9999	.2561	2	58	.0000	.7648	2
59′	.9997	.5464	1′	59′	.9999	.2490	1′	59′	.0000	6.4637	1′
88° 268°	**9.9997**	**8.5428**	**92° 272°**	**89° 269°**	**9.9999**	**8.2419**	**91° 271°**	**90° 270°**	**10.0000**		**90° 270°**

TABLE V.

HORIZONTAL DISTANCES AND ELEVATIONS FROM STADIA READINGS. § 204.

Minutes.	0°		1°		2°		3°	
	Hor. Dist.	Diff. Elev.	Hor. Dist.	Diff. Elev.	Hor. Dist.	Diff. Elev.	Hor. Dist.	Diff. Elev.
0 . .	100.00	0.00	99.97	1.74	99.88	3.49	99.73	5.23
2 . .	"	0.06	"	1.80	99.87	3.55	99.72	5.28
4 . .	"	0.12	"	1.86	"	3.60	99.71	5.34
6 . .	"	0.17	99.96	1.92	"	3.66	"	5.40
8 . .	"	0.23	"	1.98	99.86	3.72	99.70	5.46
10 . .	"	0.29	"	2.04	"	3.78	99.69	5.52
12 . .	"	0.35	"	2.09	99.85	3.84	"	5.57
14 . .	"	0.41	99.95	2.15	"	3.90	99.68	5.63
16 . .	"	0.47	"	2.21	99.84	3.95	"	5.69
18 . .	"	0.52	"	2.27	"	4.01	99.67	5.75
20 . .	"	0.58	"	2.33	99.83	4.07	99.66	5.80
22 . .	"	0.64	99.94	2.38	"	4.13	"	5.86
24 . .	"	0.70	"	2.44	99.82	4.18	99.65	5.92
26 . .	99.99	0.76	"	2.50	"	4.24	99.64	5.98
28 . .	"	0.81	99.93	2.56	99.81	4.30	99.63	6.04
30 . .	"	0.87	"	2.62	"	4.36	"	6.09
32 . .	"	0.93	"	2.67	99.80	4.42	99.62	6.15
34 . .	"	0.99	"	2.73	"	4.48	"	6.21
36 . .	"	1.05	99.92	2.79	99.79	4.53	99.61	6.27
38 . .	"	1.11	"	2.85	"	4.59	99.60	6.33
40 . .	"	1.16	"	2.91	99.78	4.65	99.59	6.38
42 . .	"	1.22	99.91	2.97	"	4.71	"	6.44
44 . .	99.98	1.28	"	3.02	99.77	4.76	99.58	6.50
46 . .	"	1.34	99.90	3.08	"	4.82	99.57	6.56
48 . .	"	1.40	"	3.14	99.76	4.88	99.56	6.61
50 . .	"	1.45	"	3.20	"	4.94	"	6.67
52 . .	"	1.51	99.89	3.26	99.75	4.99	99.55	6.73
54 . .	"	1.57	"	3.31	99.74	5.05	99.54	6.78
56 . .	99.97	1.63	"	3.37	"	5.11	99.53	6.84
58 . .	"	1.69	99.88	3.43	99.73	5.17	99.52	6.90
60 . .	"	1.74	"	3.49	"	5.23	99.51	6.96
$c = 0.75$	0.75	0.01	0.75	0.02	0.75	0.03	0.75	0.05
$c = 1.00$	1.00	0.01	1.00	0.03	1.00	0.04	1.00	0.06
$c = 1.25$	1 25	0.02	1.25	0.03	1.25	0.05	1.25	0.08

* This table was computed by Mr. Arthur Winslow of the State Geological Survey of Pennsylvania. For description of chart for graphical reduction see p. v.

TABLE V.—*Continued.*

HORIZONTAL DISTANCES AND ELEVATIONS FROM STADIA READINGS.

Minutes.	4°		5°		6°		7°	
	Hor. Dist.	Diff. Elev.	Hor. Dist.	Diff. Elev.	Hor. Dist.	Diff. Elev.	Hor. Dist.	Diff. Elev.
0 . .	99.51	6.96	99.24	8.68	98.91	10.40	98.51	12.10
2 . .	"	7.02	99.23	8.74	98.90	10.45	98.50	12.15
4 . .	99.50	7.07	99.22	8.80	98.88	10.51	98.48	12.21
6 . .	99.49	7.13	99.21	8.85	98.87	10.57	98.47	12.26
8 . .	99.48	7.19	99.20	8.91	98.86	10.62	98.46	12.32
10 . .	99.47	7.25	99.19	8.97	98.85	10.68	98.44	12.38
12 . .	99.46	7.30	99.18	9.03	98.83	10.74	98.43	12.43
14 . .	"	7.36	99.17	9.08	98.82	10.79	98.41	12.49
16 . .	99.45	7.42	99.16	9.14	98.81	10.85	98.40	12.55
18 . .	99.44	7.48	99.15	9.20	98.80	10.91	98.39	12.60
20 . .	99.43	7.53	99.14	9.25	98.78	10.96	98.37	12.66
22 . .	99.42	7.59	99.13	9.31	98.77	11.02	98.36	12.72
24 . .	99.41	7.65	99.11	9.37	98.76	11.08	98.34	12.77
26 . .	99.40	7.71	99.10	9.43	98.74	11.13	98.33	12.83
28 . .	99.39	7.76	99.09	9.48	98.73	11.19	98.31	12.88
30 . .	99.38	7.82	99.08	9.54	98.72	11.25	98.29	12.94
32 . .	99.38	7.88	99.07	9.60	98.71	11.30	98.28	13.00
34 . .	99.37	7.94	99.06	9.65	98.69	11.36	98.27	13.05
36 . .	99.36	7.99	99.05	9.71	98.68	11.42	98.25	13.11
38 . .	99.35	8.05	99.04	9.77	98.67	11.47	98.24	13.17
40 . .	99.34	8.11	99.03	9.83	98.65	11.53	98.22	13.22
42 . .	99.33	8.17	99.01	9.88	98.64	11.59	98.20	13.28
44 . .	99.32	8.22	99.00	9.94	98.63	11.64	98.19	13.33
46 . .	99.31	8.28	98.99	10.00	98.61	11.70	98.17	13.39
48 . .	99.30	8.34	98.98	10.05	98.60	11.76	98.16	13.45
50 . .	99.29	8.40	98.97	10.11	98.58	11.81	98.14	13.50
52 . .	99.28	8.45	98.96	10.17	98.57	11.87	98.13	13.56
54 . .	99.27	8.51	98.94	10.22	98.56	11.93	98.11	13.61
56 . .	99.26	8.57	98.93	10.28	98.54	11.98	98.10	13.67
58 . .	99.25	8.63	98.92	10.34	98.53	12.04	98.08	13.73
60 . .	99.24	8.68	98.91	10.40	98.51	12.10	98.06	13.78
$c = 0.75$	0.75	0.06	0.75	0.07	0.75	0.08	0.74	0.10
$c = 1.00$	1.00	0.08	0.99	0.09	0.99	0.11	0.99	0.13
$c = 1.25$	1.25	0.10	1.24	0.11	1.24	0.14	1.24	0.16

TABLE V.—*Continued.*

HORIZONTAL DISTANCES AND ELEVATIONS FROM STADIA READINGS.

Minutes.	8°		9°		10°		11°	
	Hor. Dist.	Diff. Elev.	Hor. Dist.	Diff. Elev.	Hor. Dist.	Diff. Elev.	Hor. Dist.	Diff. Elev.
0 . .	98.06	13.78	97.55	15.45	96.98	17.10	96.36	18.73
2 . .	98.05	13.84	97.53	15.51	96.96	17.16	96.34	18.78
4 . .	98.03	13.89	97.52	15.56	96.94	17.21	96.32	18.84
6 . .	98.01	13.95	97.50	15.62	96.92	17.26	96.29	18.89
8 . .	98.00	14.01	97.48	15.67	96.90	17.32	96.27	18.95
10 . .	97.98	14.06	97.46	15.73	96.88	17.37	96.25	19.00
12 . .	97.97	14.12	97.44	15.78	96.86	17.43	96.23	19.05
14 . .	97.95	14.17	97.43	15.84	96.84	17.48	96.21	19.11
16 . .	97.93	14.23	97.41	15.89	96.82	17.54	96.18	19.16
18 . .	97.92	14.28	97.39	15.95	96.80	17.59	96.16	19.21
20 . .	97.90	14.34	97.37	16.00	96.78	17.65	96.14	19.27
22 . .	97.88	14.40	97.35	16.06	96.76	17.70	96.12	19.32
24 . .	97.87	14.45	97.33	16.11	96.74	17.76	96.09	19.38
26 . .	97.85	14.51	97.31	16.17	96.72	17.81	96.07	19.43
28 . .	97.83	14.56	97.29	16.22	96.70	17.86	96.05	19.48
30 . .	97.82	14.62	97.28	16.28	96.68	17.92	96.03	19.54
32 . .	97.80	14.67	97.26	16.33	96.66	17.97	96.00	19.59
34 . .	97.78	14.73	97.24	16.39	96.64	18.03	95.98	19.64
36 . .	97.76	14.79	97.22	16.44	96.62	18.08	95.96	19.70
38 . .	97.75	14.84	97.20	16.50	96.60	18.14	95.93	19.75
40 . .	97.73	14.90	97.18	16.55	96.57	18.19	95.91	19.80
42 . .	97.71	14.95	97.16	16.61	96.55	18.24	95.89	19.86
44 . .	97.69	15.01	97.14	16.66	96.53	18.30	95.86	19.91
46 . .	97.68	15.06	97.12	16.72	96.51	18.35	95.84	19.96
48 . .	97.66	15.12	97.10	16.77	96.49	18.41	95.82	20.02
50 . .	97.64	15.17	97.08	16.83	96.47	18.46	95.79	20.07
52 . .	97.62	15.23	97.06	16.88	96.45	18.51	95.77	20.12
54 . .	97.61	15.28	97.04	16.94	96.42	18.57	95.75	20.18
56 . .	97.59	15.34	97.02	16.99	96.40	18.62	95.72	20.23
58 . .	97.57	15.40	97.00	17.05	96.38	18.68	95.70	20.28
60 . .	97.55	15.45	96.98	17.10	96.36	18.73	95.68	20.34
$c = 0.75$	0.74	0.11	0.74	0.12	0.74	0.14	0.73	0.15
$c = 1.00$	0.99	0.15	0.99	0.16	0.98	0.18	0.98	0.20
$c = 1.25$	1.23	0.18	1.23	0.21	1.23	0.23	1.22	0.25

TABLE V.—*Continued.*

HORIZONTAL DISTANCES AND ELEVATIONS FROM STADIA READINGS.

Minutes.	12°		13°		14°		15°	
	Hor. Dist.	Diff. Elev.	Hor. Dist.	Diff. Elev.	Hor. Dist.	Diff. Elev.	Hor. Dist.	Diff. Elev.
0 . .	95.68	20.34	94.94	21.92	94.15	23.47	93.30	25.00
2 . .	95.65	20.39	94.91	21.97	94.12	23.52	93.27	25.05
4 . .	95.63	20.44	94.89	22.02	94.09	23.58	93.24	25.10
6 . .	95.61	20.50	94.86	22.08	94.07	23.63	93.21	25.15
8 . .	95.58	20.55	94.84	22.13	94.04	23.68	93.18	25.20
10 . .	95.56	20.60	94.81	22.18	94.01	23.73	93.16	25.25
12 . .	95.53	20.66	94.79	22.23	93.98	23.78	93.13	25.30
14 . .	95.51	20.71	94.76	22.28	93.95	23.83	93.10	25.35
16 . .	95.49	20.76	94.73	22.34	93.93	23.88	93.07	25.40
18 . .	95.46	20.81	94.71	22.39	93.90	23.93	93.04	25.45
20 . .	95.44	20.87	94.68	22.44	93.87	23.99	93.01	25.50
22 . .	95.41	20.92	94.66	22.49	93.84	24.04	92.98	25.55
24 . .	95.39	20.97	94.63	22.54	93.81	24.09	92.95	25.60
26 . .	95.36	21.03	94.60	22.60	93.79	24.14	92.92	25.65
28 . .	95.34	21.08	94.58	22.65	93.76	24.19	92.89	25.70
30 . .	95.32	21.13	94.55	22.70	93.73	24.24	92.86	25.75
32 . .	95.29	21.18	94.52	22.75	93.70	24.29	92.83	25.80
34 . .	95.27	21.24	94.50	22.80	93.67	24.34	92.80	25.85
36 . .	95.24	21.29	94.47	22.85	93.65	24.39	92.77	25.90
38 . .	95.22	21.34	94.44	22.91	93.62	24.44	92.74	25.95
40 . .	95.19	21.39	94.42	22.96	93.59	24.49	92.71	26.00
42 . .	95.17	21.45	94.39	23.01	93.56	24.55	92.68	26.05
44 . .	95.14	21.50	94.36	23.06	93.53	24.60	92.65	26.10
46 . .	95.12	21.55	94.34	23.11	93.50	24.65	92.62	26.15
48 . .	95.09	21.60	94.31	23.16	93.47	24.70	92.59	26.20
50 . .	95.07	21.66	94.28	23.22	93.45	24.75	92.56	26.25
52 . .	95.04	21.71	94.26	23.27	93.42	24.80	92.53	26.30
54 . .	95.02	21.76	94.23	23.32	93.39	24.85	92.49	26.35
56 . .	94.99	21.81	94.20	23.37	93.36	24.90	92.46	26.40
58 . .	94.97	21.87	94.17	23.42	93.33	24.95	92.43	26.45
60 . .	94.94	21.92	94.15	23.47	93.30	25.00	92.40	26.50
$c = 0.75$	0.73	0.16	0.73	0.17	0.73	0.19	0.72	0.20
$c = 1.00$	0.98	0.22	0.97	0.23	0.97	0.25	0.96	0.27
$c = 1.25$	1.22	0.27	1.21	0.29	1.21	0.31	1.20	0.34

TABLE V.—*Continued.*

HORIZONTAL DISTANCES AND ELEVATIONS FROM STADIA READINGS.

Minutes.	16°		17°		18°		19°	
	Hor. Dist.	Diff. Elev.	Hor. Dist.	Diff. Elev.	Hor. Dist.	Diff. Elev.	Hor. Dist.	Diff. Elev.
0 . .	92.40	26.50	91.45	27.96	90.45	29.39	89.40	30.78
2 . .	92.37	26.55	91.42	28.01	90.42	29.44	89.36	30.83
4 . .	92.34	26.59	91.39	28.06	90.38	29.48	89.33	30.87
6 . .	92.31	26.64	91.35	28.10	90.35	29.53	89.29	30.92
8 . .	92.28	26.69	91.32	28.15	90.31	29.58	89.26	30.97
10 . .	92.25	26.74	91.29	28.20	90.28	29.62	89.22	31.01
12 . .	92.22	26.79	91.26	28.25	90.24	29.67	89.18	31.06
14 . .	92.19	26.84	91.22	28.30	90.21	29.72	89.15	31.10
16 . .	92.15	26.89	91.19	28.34	90.18	29.76	89.11	31.15
18 . .	92.12	26.94	91.16	28.39	90.14	29.81	89.08	31.19
20 . .	92.09	26.99	91.12	28.44	90.11	29.86	89.04	31.24
22 . .	92.06	27.04	91.09	28.49	90.07	29.90	89.00	31.28
24 . .	92.03	27.09	91.06	28.54	90.04	29.95	88.96	31.33
26 . .	92.00	27.13	91.02	28.58	90.00	30.00	88.93	31.38
28 . .	91.97	27.18	90.99	28.63	89.97	30.04	88.89	31.42
30 . .	91.93	27.23	90.96	28.68	89.93	30.09	88.86	31.47
32 . .	91.90	27.28	90.92	28.73	89.90	30.14	88.82	31.51
34 . .	91.87	27.33	90.89	28.77	89.86	30.19	88.78	31.56
36 . .	91.84	27.38	90.86	28.82	89.83	30.23	88.75	31.60
38 . .	91.81	27.43	90.82	28.87	89.79	30.28	88.71	31.65
40 . .	91.77	27.48	90.79	28.92	89.76	30.32	88.67	31.69
42 . .	91.74	27.52	90.76	28.96	89.72	30.37	88.64	31.74
44 . .	91.71	27.57	90.72	29.01	89.69	30.41	88.60	31.78
46 . .	91.68	27.62	90.69	29.06	89.65	30.46	88.56	31.83
48 . .	91.65	27.67	90.66	29.11	89.61	30.51	88.53	31.87
50 . .	91.61	27.72	90.62	29.15	89.58	30.55	88.49	31.92
52 . .	91.58	27.77	90.59	29.20	89.54	30.60	88.45	31.96
54 . .	91.55	27.81	90.55	29.25	89.51	30.65	88.41	32.01
56 . .	91.52	27.86	90.52	29.30	89.47	30.69	88.38	32.05
58 . .	91.48	27.91	90.48	29.34	89.44	30.74	88.34	32.09
60 . .	91.45	27.96	90.45	29.39	89.40	30.78	88.30	32.14
$c = 0.75$	0.72	0.21	0.72	0.23	0.71	0.24	0.71	0.25
$c = 1.00$	0.86	0.28	0.95	0.30	0.95	0.32	0.94	0.33
$c = 1.25$	1.20	0.35	1.19	0.38	1.19	0.40	1.18	0.42

TABLE V.—*Continued.*

HORIZONTAL DISTANCES AND ELEVATIONS FROM STADIA READINGS.

Minutes.	20°		21°		22°		23°	
	Hor. Dist.	Diff. Elev.	Hor. Dist.	Diff. Elev.	Hor. Dist.	Diff. Elev.	Hor. Dist.	Diff. Elev.
0 . .	88.30	32.14	87.16	33.46	85.97	34.73	84.73	35.97
2 . .	88.26	32.18	87.12	33.50	85.93	34.77	84.69	36.01
4 . .	88.23	32.23	87.08	33.54	85.89	34.82	84.65	36.05
6 . .	88.19	32.27	87.04	33.59	85.85	34.86	84.61	36.09
8 . .	88.15	32.32	87.00	33.63	85.80	34.90	84.57	36.13
10 . .	88.11	32.36	86.96	33.67	85.76	34.94	84.52	36.17
12 . .	88.08	32.41	86.92	33.72	85.72	34.98	84.48	36.21
14 . .	88.04	32.45	86.88	33.76	85.68	35.02	84.44	36.25
16 . .	88.00	32.49	86.84	33.80	85.64	35.07	84.40	36.29
18 . .	87.96	32.54	86.80	33.84	85.60	35.11	84.35	36.33
20 . .	87.93	32.58	86.77	33.89	85.56	35.15	84.31	36.37
22 . .	87.89	32.63	86.73	33.93	85.52	35.19	84.27	36.41
24 . .	87.85	32.67	86.69	33.97	85.48	35.23	84.23	36.45
26 . .	87.81	32.72	86.65	34.01	85.44	35.27	84.18	36.49
28 . .	87.77	32.76	86.61	34.06	85.40	35.31	84.14	36.53
30 . .	87.74	32.80	86.57	34.10	85.36	35.36	84.10	36.57
32 . .	87.70	32.85	86.53	34.14	85.31	35.40	84.06	36.61
34 . .	87.66	32.89	86.49	34.18	85.27	35.44	84.01	36.65
36 . .	87.62	32.93	86.45	34.23	85.23	35.48	83.97	36.69
38 . .	87.58	32.98	86.41	34.27	85.19	35.52	83.93	36.73
40 . .	87.54	33.02	86.37	34.31	85.15	35.56	83.89	36.77
42 . .	87.51	33.07	86.33	34.35	85.11	35.60	83.84	36.80
44 . .	87.47	33.11	86.29	34.40	85.07	35.64	83.80	36.84
46 . .	87.43	33.15	86.25	34.44	85.02	35.68	83.76	36.88
48 . .	87.39	33.20	86.21	34.48	84.98	35.72	83.72	36.92
50 . .	87.35	33.24	86.17	34.52	84.94	35.76	83.67	36.96
52 . .	87.31	33.28	86.13	34.57	84.90	35.80	83.63	37.00
54 . .	87.27	33.33	86.09	34.61	84.86	35.85	83.59	37.04
56 . .	87.24	33.37	86.05	34.65	84.82	35.89	83.54	37.08
58 . .	87.20	33.41	86.01	34.69	84.77	35.93	83.50	37.12
60 . .	87.16	33.46	85.97	34.73	84.73	35.97	83.46	37.16
$c = 0.75$	0.70	0.26	0.70	0.27	0.69	0.29	0.69	0.30
$c = 1.00$	0.94	0.35	0.93	0.37	0.92	0.38	0.92	0.40
$c = 1.25$	1.17	0.44	1.16	0.46	1.15	0.48	1.15	0.50

TABLE V.—*Continued.*

HORIZONTAL DISTANCES AND ELEVATIONS FROM STADIA READINGS.

Minutes.	24°		25°		26°		27°	
	Hor. Dist.	Diff. Elev.	Hor. Dist.	Diff. Elev.	Hor. Dist.	Diff. Elev	Hor. Dist.	Diff. Elev.
0 . .	83.46	37.16	82.14	38.30	80.78	39.40	79.39	40.45
2 . .	83.41	37.20	82.09	38.34	80.74	39.44	79.34	40.49
4 . .	83.37	37.23	82.05	38.38	80.69	39.47	79.30	40.52
6 . .	83.33	37.27	82.01	38.41	80.65	39.51	79.25	40.55
8 . .	83.28	37.31	81.96	38.45	80.60	39.54	79.20	40.59
10 . .	83.24	37.35	81.92	38.49	80.55	39.58	79.15	40.62
12 . .	83.20	37.39	81.87	38.53	80.51	39.61	79.11	40.66
14 . .	83.15	37.43	81.83	38.56	80.46	39.65	79.06	40.69
16 . .	83.11	37.47	81.78	38.60	80.41	39.69	79.01	40.72
18 . .	83.07	37.51	81.74	38.64	80.37	39.72	78.96	40.76
20 . .	83.02	37.54	81.69	38.67	80.32	39.76	78.92	40.79
22 . .	82.98	37.58	81.65	38.71	80.28	39.79	78.87	40.82
24 . .	82.93	37.62	81.60	38.75	80.23	39.83	78.82	40.86
26 . .	82.89	37.66	81.56	38.78	80.18	39.86	78.77	40.89
28 . .	82.85	37.70	81.51	38.62	80.14	39.90	78.73	40.92
30 . .	82.80	37.74	81.47	38.86	80.09	39.93	78.68	40.96
32 . .	82.76	37.77	81.42	38.89	80.04	39.97	78.63	40.99
34 . .	82.72	37.81	81.38	38.93	80.00	40.00	78.58	41.02
36 . .	82.67	37.85	81.33	38.97	79.95	40.04	78.54	41.06
38 . .	82.63	37.89	81.28	39.00	79.90	40.07	78.49	41.09
40 . .	82.58	37.93	81.24	39.04	79.86	40.11	78.44	41.12
42 . .	82.54	37.96	81.19	39.08	79.81	40.14	78.39	41.16
44 . .	82.49	38.00	81.15	39.11	79.76	40.18	78.34	41.19
46 . .	82.45	38.04	81.10	39.15	79.72	40.21	78.30	41.22
48 . .	82.41	38.08	81.06	39.18	79.67	40.24	78.25	41.26
50 . .	82.36	38.11	81.01	39.22	79.62	40.28	78.20	41.29
52 . .	82.32	38.15	80.97	39.26	79.58	40.31	78.15	41.32
54 . .	82.27	38.19	80.92	39.29	79.53	40.35	78.10	41.35
56 . .	82.23	38.23	80.87	39.33	79.48	40.38	78.06	41.39
58 . .	82.18	38.26	80.83	39.36	79.44	40.42	78.01	41.42
60 . .	82.14	38.30	80.78	39.40	79.39	40.45	77.96	41.45
$c = 0.75$	0.68	0.31	0.68	0.32	0.67	0.33	0.66	0.35
$c = 1.00$	0.91	0.41	0.90	0.43	0.89	0.45	0.89	0.46
$c = 1.25$	1.14	0.52	1.13	0.54	1.12	0.56	1.11	0.58

TABLE V.—*Continued.*

HORIZONTAL DISTANCES AND ELEVATIONS FROM STADIA READINGS.

Minutes.	28°		29°		30°	
	Hor. Dist.	Diff. Elev.	Hor. Dist	Diff. Elev.	Hor. Dist.	Diff. Elev.
0 . .	77.96	41.45	76.50	42.40	75.00	43.30
2 . .	77.91	41.48	76.45	42.43	74.95	43.33
4 . .	77.86	41.52	76.40	42.46	74.90	43.36
6 . .	77.81	41.55	76.35	42.49	74.85	43.39
8 . .	77.77	41.58	76.30	42.53	74.80	43.42
10 . .	77.72	41.61	76.25	42.56	74.75	43.45
12 . .	77.67	41.65	76.20	42.59	74.70	43.47
14 . .	77.62	41.68	76.15	42.62	74.65	43.50
16 . .	77.57	41.71	76.10	42.65	74.60	43.53
18 . .	77.52	41.74	76.05	42.68	74.55	43.56
20 . .	77.48	41.77	76.00	42.71	74.49	43.59
22 . .	77.42	41.81	75.95	42.74	74.44	43.62
24 . .	77.38	41.84	75.90	42.77	74.39	43.65
26 . .	77.33	41.87	75.85	42.80	74.34	43.67
28 . .	77.28	41.90	75.80	42.83	74.29	43.70
30 . .	77.23	41.93	75.75	42.86	74.24	43.73
32 . .	77.18	41.97	75.70	42.89	74.19	43.76
34 . .	77.13	42.00	75.65	42.92	74.14	43.79
36 . .	77.09	42.03	75.60	42.95	74.09	43.82
38 . .	77.04	42.06	75.55	42.98	74.04	43.84
40 . .	76.99	42.09	75.50	43.01	73.99	43.87
42 . .	76.94	42.12	75.45	43.04	73.93	43.90
44 . .	76.89	42.15	75.40	43.07	73.88	43.93
46 . .	76.84	42.19	75.35	43.10	73.83	43.95
48 . .	76.79	42.22	75.30	43.13	73.78	43.98
50 . .	76.74	42.25	75.25	43.16	73.73	44.01
52 . .	76.69	42.28	75.20	43.18	73.68	44.04
54 . .	76.64	42.31	75.15	43.21	73.63	44.07
56 . .	76.59	42.34	75.10	43.24	73.58	44.09
58 . .	76.55	42.37	75.05	43.27	73.52	44.12
60 . .	76.50	42.40	75.00	43.30	73.47	44.15
$c = 0.75$	0.66	0.36	0.65	0.37	0.65	0.38
$c = 1.00$	0.88	0.48	0.87	0.49	0.86	0.51
$c = 1.25$	1.10	0.60	1.09	0.62	1.08	0.64

TABLE VI.

Natural Sines and Cosines.

′	0° Sine	0° Cosin	1° Sine	1° Cosin	2° Sine	2° Cosin	3° Sine	3° Cosin	4° Sine	4° Cosin	′
0	.00000	One.	.01745	.99985	.03490	.99939	.05234	.99863	.06976	.99756	60
1	.00029	One.	.01774	.99984	.03519	.99938	.05263	.99861	.07005	.99754	59
2	.00058	One.	.01803	.99984	.03548	.99937	.05292	.99860	.07034	.99752	58
3	.00087	One.	.01832	.99983	.03577	.99936	.05321	.99858	.07063	.99750	57
4	.00116	One.	.01862	.99983	.03606	.99935	.05350	.99857	.07092	.99748	56
5	.00145	One.	.01891	.99982	.03635	.99934	.05379	.99855	.07121	.99746	55
6	.00175	One.	.01920	.99982	.03664	.99933	.05408	.99854	.07150	.99744	54
7	.00204	One.	.01949	.99981	.03693	.99932	.05437	.99852	.07179	.99742	53
8	.00233	One.	.01978	.99980	.03723	.99931	.05466	.99851	.07208	.99740	52
9	.00262	One.	.02007	.99980	.03752	.99930	.05495	.99849	.07237	.99738	51
10	.00291	One.	.02036	.99979	.03781	.99929	.05524	.99847	.07266	.99736	50
11	.00320	.99999	.02065	.99979	.03810	.99927	.05553	.99846	.07295	.99734	49
12	.00349	.99999	.02094	.99978	.03839	.99926	.05582	.99844	.07324	.99731	48
13	.00378	.99999	.02123	.99977	.03868	.99925	.05611	.99842	.07353	.99729	47
14	.00407	.99999	.02152	.99977	.03897	.99924	.05640	.99841	.07382	.99727	46
15	.00436	.99999	.02181	.99976	.03926	.99923	.05669	.99839	.07411	.99725	45
16	.00465	.99999	.02211	.99976	.03955	.99922	.05698	.99838	.07440	.99723	44
17	.00495	.99999	.02240	.99975	.03984	.99921	.05727	.99836	.07469	.99721	43
18	.00524	.99999	.02269	.99974	.04013	.99919	.05756	.99834	.07498	.99719	42
19	.00553	.99998	.02298	.99974	.04042	.99918	.05785	.99833	.07527	.99716	41
20	.00582	.99998	.02327	.99973	.04071	.99917	.05814	.99831	.07556	.99714	40
21	.00611	.99998	.02356	.99972	.04100	.99916	.05944	.99829	.07585	.99712	39
22	.00640	.99998	.02385	.99972	.04129	.99915	.05873	.99827	.07614	.99710	38
23	.00669	.99998	.02414	.99971	.04159	.99913	.05902	.99826	.07643	.99708	37
24	.00698	.99998	.02443	.99970	.04188	.99912	.05931	.99824	.07672	.99705	36
25	.00727	.99997	.02472	.99969	.04217	.99911	.05960	.99822	.07701	.99703	35
26	.00756	.99997	.02501	.99969	.04246	.99910	.05989	.99821	.07730	.99701	34
27	.00785	.99997	.02530	.99968	.04275	.99909	.06018	.99819	.07759	.99699	33
28	.00814	.99997	.02560	.99967	.04304	.99907	.06047	.99817	.07788	.99696	32
29	.00844	.99996	.02589	.99966	.04333	.99906	.06076	.99815	.07817	.99694	31
30	.00873	.99996	.02618	.99966	.04362	.99905	.06105	.99813	.07846	.99692	30
31	.00902	.99996	.02647	.99965	.04391	.99904	.06134	.99812	.07875	.99689	29
32	.00931	.99996	.02676	.99964	.04420	.99902	.06163	.99810	.07904	.99687	28
33	.00960	.99995	.02705	.99963	.04449	.99901	.06192	.99808	.07933	.99685	27
34	.00989	.99995	.02734	.99963	.04478	.99900	.06221	.99806	.07962	.99683	26
35	.01018	.99995	.02763	.99962	.04507	.99898	.06250	.99804	.07991	.99680	25
36	.01047	.99995	.02792	.99961	.04536	.99897	.06279	.99803	.08020	.99678	24
37	.01076	.99994	.02821	.99960	.04565	.99896	.06308	.99801	.08049	.99676	23
38	.01105	.99994	.02850	.99959	.04594	.99894	.06337	.99799	.08078	.99673	22
39	.01134	.99994	.02879	.99959	.04623	.99893	.06366	.99797	.08107	.99671	21
40	.01164	.99993	.02908	.99958	.04653	.99892	.06395	.99795	.08136	.99668	20
41	.01193	.99993	.02938	.99957	.04682	.99890	.06424	.99793	.08165	.99666	19
42	.01222	.99993	.02967	.99956	.04711	.99889	.06453	.99792	.08194	.99664	18
43	.01251	.99992	.02996	.99955	.04740	.99888	.06482	.99790	.08223	.99661	17
44	.01280	.99992	.03025	.99954	.04769	.99886	.06511	.99788	.08252	.99659	16
45	.01309	.99991	.03054	.99953	.04798	.99885	.06540	.99786	.08281	.99657	15
46	.01338	.99991	.03083	.99952	.04827	.99883	.06569	.99784	.08310	.99654	14
47	.01367	.99991	.03112	.99952	.04856	.99882	.06598	.99782	.08339	.99652	13
48	.01396	.99990	.03141	.99951	.04885	.99881	.06627	.99780	.08368	.99649	12
49	.01425	.99990	.03170	.99950	.04914	.99879	.06656	.99778	.08397	.99647	11
50	.01454	.99989	.03199	.99949	.04943	.99878	.06685	.99776	.08426	.99644	10
51	.01483	.99989	.03228	.99948	.04972	.99876	.06714	.99774	.08455	.99642	9
52	.01513	.99989	.03257	.99947	.05001	.99875	.06743	.99772	.08484	.99639	8
53	.01542	.99988	.03286	.99946	.05030	.99873	.06773	.99770	.08513	.99637	7
54	.01571	.99988	.03316	.99945	.05059	.99872	.06802	.99768	.08542	.99635	6
55	.01600	.99987	.03345	.99944	.05088	.99870	.06831	.99766	.08571	.99632	5
56	.01629	.99987	.03374	.99943	.05117	.99869	.06860	.99764	.08600	.99630	4
57	.01658	.99986	.03403	.99942	.05146	.99867	.06889	.99762	.08629	.99627	3
58	.01687	.99986	.03432	.99941	.05175	.99866	.06918	.99760	.08658	.99625	2
59	.01716	.99985	.03461	.99940	.05205	.99864	.06947	.99758	.08687	.99622	1
60	.01745	.99985	.03490	.99939	.05234	.99863	.06976	.99756	.08716	.99619	0
′	Cosin	Sine	Cosin	Sine	Cosin	Sine	Cosin	Sine	Cosin	Sine	′
	89°		88°		87°		86°		85°		

TABLE VI.—*Continued.*

NATURAL SINES AND COSINES.

′	5° Sine	5° Cosin	6° Sine	6° Cosin	7° Sine	7° Cosin	8° Sine	8° Cosin	9° Sine	9° Cosin	′
0	.08716	.99619	.10453	.99452	.12187	.99255	.13917	.99027	.15643	.98769	60
1	.08745	.99617	.10482	.99449	.12216	.99251	.13946	.99023	.15672	.98764	59
2	.08774	.99614	.10511	.99446	.12245	.99248	.13975	.99019	.15701	.98760	58
3	.08803	.99612	.10540	.99443	.12274	.99244	.14004	.99015	.15730	.98755	57
4	.08831	.99609	.10569	.99440	.12302	.99240	.14033	.99011	.15758	.98751	56
5	.08860	.99607	.10597	.99437	.12331	.99237	.14061	.99006	.15787	.98746	55
6	.08889	.99604	.10626	.99434	.12360	.99233	.14090	.99002	.15816	.98741	54
7	.08918	.99602	.10655	.99431	.12389	.99230	.14119	.98998	.15845	.98737	53
8	.08947	.99599	.10684	.99428	.12418	.99226	.14148	.98994	.15873	.98732	52
9	.08976	.99596	.10713	.99424	.12447	.99222	.14177	.98990	.15902	.98728	51
10	.09005	.99594	.10742	.99421	.12476	.99219	.14205	.98986	.15931	.98723	50
11	.09034	.99591	.10771	.99418	.12504	.99215	.14234	.98982	.15959	.98718	49
12	.09063	.99588	.10800	.99415	.12533	.99211	.14263	.98978	.15988	.98714	48
13	.09092	.99586	.10829	.99412	.12562	.99208	.14292	.98973	.16017	.98709	47
14	.09121	.99583	.10858	.99409	.12591	.99204	.14320	.98969	.16046	.98704	46
15	.09150	.99580	.10887	.99406	.12620	.99200	.14349	.98965	.16074	.98700	45
16	.09179	.99578	.10916	.99402	.12649	.99197	.14378	.98961	.16103	.98695	44
17	.09208	.99575	.10945	.99399	.12678	.99193	.14407	.98957	.16132	.98690	43
18	.09237	.99572	.10973	.99396	.12706	.99189	.14436	.98953	.16160	.98686	42
19	.09266	.99570	.11002	.99393	.12735	.99186	.14464	.98948	.16189	.98681	41
20	.09295	.99567	.11031	.99390	.12764	.99182	.14493	.98944	.16218	.98676	40
21	.09324	.99564	.11060	.99386	.12793	.99178	.14522	.98940	.16246	.98671	39
22	.09353	.99562	.11089	.99383	.12822	.99175	.14551	.98936	.16275	.98667	38
23	.09382	.99559	.11118	.99380	.12851	.99171	.14580	.98931	.16304	.98662	37
24	.09411	.99556	.11147	.99377	.12880	.99167	.14608	.98927	.16333	.98657	36
25	.09440	.99553	.11176	.99374	.12908	.99163	.14637	.98923	.16361	.98652	35
26	.09469	.99551	.11205	.99370	.12937	.99160	.14666	.98919	.16390	.98648	34
27	.09498	.99548	.11234	.99367	.12966	.99156	.14695	.98914	.16419	.98643	33
28	.09527	.99545	.11263	.99364	.12995	.99152	.14723	.98910	.16447	.98638	32
29	.09556	.99542	.11291	.99360	.13024	.99148	.14752	.98906	.16476	.98633	31
30	.09585	.99540	.11320	.99357	.13053	.99144	.14781	.98902	.16505	.98629	30
31	.09614	.99537	.11349	.99354	.13081	.99141	.14810	.98897	.16533	.98624	29
32	.09642	.99534	.11378	.99351	.13110	.99137	.14838	.98893	.16562	.98619	28
33	.09671	.99531	.11407	.99347	.13139	.99133	.14867	.98889	.16591	.98614	27
34	.09700	.99528	.11436	.99344	.13168	.99129	.14896	.98884	.16620	.98609	26
35	.09729	.99526	.11465	.99341	.13197	.99125	.14925	.98880	.16648	.98604	25
36	.09758	.99523	.11494	.99337	.13226	.99122	.14954	.98876	.16677	.98600	24
37	.09787	.99520	.11523	.99334	.13254	.99118	.14982	.98871	.16706	.98595	23
38	.09816	.99517	.11552	.99331	.13283	.99114	.15011	.98867	.16734	.98590	22
39	.09845	.99514	.11580	.99327	.13312	.99110	.15040	.98863	.16763	.98585	21
40	.09874	.99511	.11609	.99324	.13341	.99106	.15069	.98858	.16792	.98580	20
41	.09903	.99508	.11638	.99320	.13370	.99102	.15097	.98854	.16820	.98575	19
42	.09932	.99506	.11667	.99317	.13399	.99098	.15126	.98849	.16849	.98570	18
43	.09961	.99503	.11696	.99314	.13427	.99094	.15155	.98845	.16878	.98565	17
44	.09990	.99500	.11725	.99310	.13456	.99091	.15184	.98841	.16906	.98561	16
45	.10019	.99497	.11754	.99307	.13485	.99087	.15212	.98836	.16935	.98556	15
46	.10048	.99494	.11783	.99303	.13514	.99083	.15241	.98832	.16964	.98551	14
47	.10077	.99491	.11812	.99300	.13543	.99079	.15270	.98827	.16992	.98546	13
48	.10106	.99488	.11840	.99297	.13572	.99075	.15299	.98823	.17021	.98541	12
49	.10135	.99485	.11869	.99293	.13600	.99071	.15327	.98818	.17050	.98536	11
50	.10164	.99482	.11898	.99290	.13629	.99067	.15356	.98814	.17078	.98531	10
51	.10192	.99479	.11927	.99286	.13658	.99063	.15385	.98809	.17107	.98526	9
52	.10221	.99476	.11956	.99283	.13687	.99059	.15414	.98805	.17136	.98521	8
53	.10250	.99473	.11985	.99279	.13716	.99055	.15442	.98800	.17164	.98516	7
54	.10279	.99470	.12014	.99276	.13744	.99051	.15471	.98796	.17193	.98511	6
55	.10308	.99467	.12043	.99272	.13773	.99047	.15500	.98791	.17222	.98506	5
56	.10337	.99464	.12071	.99269	.13802	.99043	.15529	.98787	.17250	.98501	4
57	.10366	.99461	.12100	.99265	.13831	.99039	.15557	.98782	.17279	.98496	3
58	.10395	.99458	.12129	.99262	.13860	.99035	.15586	.98778	.17308	.98491	2
59	.10424	.99455	.12158	.99258	.13889	.99031	.15615	.98773	.17336	.98486	1
60	.10453	.99452	.12187	.99255	.13917	.99027	.15643	.98769	.17365	.98481	0
′	84° Cosin	84° Sine	83° Cosin	83° Sine	82° Cosin	82° Sine	81° Cosin	81° Sine	80° Cosin	80° Sine	′

TABLE VI.—*Continued.*

NATURAL SINES AND COSINES.

′	10° Sine	10° Cosin	11° Sine	11° Cosin	12° Sine	12° Cosin	13° Sine	13° Cosin	14° Sine	14° Cosin	′
0	.17365	.98481	.19081	.98163	.20791	.97815	.22495	.97437	.24192	.97030	60
1	.17393	.98476	.19109	.98157	.20820	.97809	.22523	.97430	.24220	.97023	59
2	.17422	.98471	.19138	.98152	.20848	.97803	.22552	.97424	.24249	.97015	58
3	.17451	.98466	.19167	.98146	.20877	.97797	.22580	.97417	.24277	.97008	57
4	.17479	.98461	.19195	.98140	.20905	.97791	.22608	.97411	.24305	.97001	56
5	.17508	.98455	.19224	.98135	.20933	.97784	.22637	.97404	.24333	.96994	55
6	.17537	.98450	.19252	.98129	.20962	.97778	.22665	.97398	.24362	.96987	54
7	.17565	.98445	.19281	.98124	.20990	.97772	.22693	.97391	.24390	.96980	53
8	.17594	.98440	.19309	.98118	.21019	.97766	.22722	.97384	.24418	.96973	52
9	.17623	.98435	.19338	.98112	.21047	.97760	.22750	.97378	.24446	.96966	51
10	.17651	.98430	.19366	.98107	.21076	.97754	.22778	.97371	.24474	.96959	50
11	.17680	.98425	.19395	.98101	.21104	.97748	.22807	.97365	.24503	.96952	49
12	.17708	.98420	.19423	.98096	.21132	.97742	.22835	.97358	.24531	.96945	48
13	.17737	.98414	.19452	.98090	.21161	.97735	.22863	.97351	.24559	.96937	47
14	.17766	.98409	.19481	.98084	.21189	.97729	.22892	.97345	.24587	.96930	46
15	.17794	.98404	.19509	.98079	.21218	.97723	.22920	.97338	.24615	.96923	45
16	.17823	.98399	.19538	.98073	.21246	.97717	.22948	.97331	.24644	.96916	44
17	.17852	.98394	.19566	.98067	.21275	.97711	.22977	.97325	.24672	.96909	43
18	.17880	.98389	.19595	.98061	.21303	.97705	.23005	.97318	.24700	.96902	42
19	.17909	.98383	.19623	.98056	.21331	.97698	.23033	.97311	.24728	.96894	41
20	.17937	.98378	.19652	.98050	.21360	.97692	.23062	.97304	.24756	.96887	40
21	.17966	.98373	.19680	.98044	.21388	.97686	.23090	.97298	.24784	.96880	39
22	.17995	.98368	.19709	.98039	.21417	.97680	.23118	.97291	.24813	.96873	38
23	.18023	.98362	.19737	.98033	.21445	.97673	.23146	.97284	.24841	.96866	37
24	.18052	.98357	.19766	.98027	.21474	.97667	.23175	.97278	.24869	.96858	36
25	.18081	.98352	.19794	.98021	.21502	.97661	.23203	.97271	.24897	.96851	35
26	.18109	.98347	.19823	.98016	.21530	.97655	.23231	.97264	.24925	.96844	34
27	.18138	.98341	.19851	.98010	.21559	.97648	.23260	.97257	.24954	.96837	33
28	.18166	.98336	.19880	.98004	.21587	.97642	.23288	.97251	.24982	.96829	32
29	.18195	.98331	.19908	.97998	.21616	.97636	.23316	.97244	.25010	.96822	31
30	.18224	.98325	.19937	.97992	.21644	.97630	.23345	.97237	.25038	.96815	30
31	.18252	.98320	.19965	.97987	.21672	.97623	.23373	.97230	.25066	.96807	29
32	.18281	.98315	.19994	.97981	.21701	.97617	.23401	.97223	.25094	.96800	28
33	.18309	.98310	.20022	.97975	.21729	.97611	.23429	.97217	.25122	.96793	27
34	.18338	.98304	.20051	.97969	.21758	.97604	.23458	.97210	.25151	.96786	26
35	.18367	.98299	.20079	.97963	.21786	.97598	.23486	.97203	.25179	.96778	25
36	.18395	.98294	.20108	.97958	.21814	.97592	.23514	.97196	.25207	.96771	24
37	.18424	.98288	.20136	.97952	.21843	.97585	.23542	.97189	.25235	.96764	23
38	.18452	.98283	.20165	.97946	.21871	.97579	.23571	.97182	.25263	.96756	22
39	.18481	.98277	.20193	.97940	.21899	.97573	.23599	.97176	.25291	.96749	21
40	.18509	.98272	.20222	.97934	.21928	.97566	.23627	.97169	.25320	.96742	20
41	.18538	.98267	.20250	.97928	.21956	.97560	.23656	.97162	.25348	.96734	19
42	.18567	.98261	.20279	.97922	.21985	.97553	.23684	.97155	.25376	.96727	18
43	.18595	.98256	.20307	.97916	.22013	.97547	.23712	.97148	.25404	.96719	17
44	.18624	.98250	.20336	.97910	.22041	.97541	.23740	.97141	.25432	.96712	16
45	.18652	.98245	.20364	.97905	.22070	.97534	.23769	.97134	.25460	.96705	15
46	.18681	.98240	.20393	.97899	.22098	.97528	.23797	.97127	.25488	.96697	14
47	.18710	.98234	.20421	.97893	.22126	.97521	.23825	.97120	.25516	.96690	13
48	.18738	.98229	.20450	.97887	.22155	.97515	.23853	.97113	.25545	.96682	12
49	.18767	.98223	.20478	.97881	.22183	.97508	.23882	.97106	.25573	.96675	11
50	.18795	.98218	.20507	.97875	.22212	.97502	.23910	.97100	.25601	.96667	10
51	.18824	.98212	.20535	.97869	.22240	.97496	.23938	.97093	.25629	.96660	9
52	.18852	.98207	.20563	.97863	.22268	.97489	.23966	.97086	.25657	.96653	8
53	.18881	.98201	.20592	.97857	.22297	.97483	.23995	.97079	.25685	.96645	7
54	.18910	.98196	.20620	.97851	.22325	.97476	.24023	.97072	.25713	.96638	6
55	.18938	.98190	.20649	.97845	.22353	.97470	.24051	.97065	.25741	.96630	5
56	.18967	.98185	.20677	.97839	.22382	.97463	.24079	.97058	.25769	.96623	4
57	.18995	.98179	.20706	.97833	.22410	.97457	.24108	.97051	.25798	.96615	3
58	.19024	.98174	.20734	.97827	.22438	.97450	.24136	.97044	.25826	.96608	2
59	.19052	.98168	.20763	.97821	.22467	.97444	.24164	.97037	.25854	.96600	1
60	.19081	.98163	.20791	.97815	.22495	.97437	.24192	.97030	.25882	.96593	0
′	Cosin	Sine	Cosin	Sine	Cosin	Sine	Cosin	Sine	Cosin	Sine	′
	79°		78°		77°		76°		75°		

TABLE VI.—*Continued.*

Natural Sines and Cosines.

′	15° Sine	15° Cosin	16° Sine	16° Cosin	17° Sine	17° Cosin	18° Sine	18° Cosin	19° Sine	19° Cosin	′
0	.25882	.96593	.27564	.96126	.29237	.95630	.30902	.95106	.32557	.94552	60
1	.25910	.96585	.27592	.96118	.29265	.95622	.30929	.95097	.32584	.94542	59
2	.25938	.96578	.27620	.96110	.29293	.95613	.30957	.95088	.32612	.94533	58
3	.25966	.96570	.27648	.96102	.29321	.95605	.30985	.95079	.32639	.94523	57
4	.25994	.96562	.27676	.96094	.29348	.95596	.31012	.95070	.32667	.94514	56
5	.26022	.96555	.27704	.96086	.29376	.95588	.31040	.95061	.32694	.94504	55
6	.26050	.96547	.27731	.96078	.29404	.95579	.31068	.95052	.32722	.94495	54
7	.26079	.96540	.27759	.96070	.29432	.95571	.31095	.95043	.32749	.94485	53
8	.26107	.96532	.27787	.96062	.29460	.95562	.31123	.95033	.32777	.94476	52
9	.26135	.96524	.27815	.96054	.29487	.95554	.31151	.95024	.32804	.94466	51
10	.26163	.96517	.27843	.96046	.29515	.95545	.31178	.95015	.32832	.94457	50
11	.26191	.96509	.27871	.96037	.29543	.95536	.31206	.95006	.32859	.94447	49
12	.26219	.96502	.27899	.96029	.29571	.95528	.31233	.94997	.32887	.94438	48
13	.26247	.96494	.27927	.96021	.29599	.95519	.31261	.94988	.32914	.94428	47
14	.26275	.96486	.27955	.96013	.29626	.95511	.31289	.94979	.32942	.94418	46
15	.26303	.96479	.27983	.96005	.29654	.95502	.31316	.94970	.32969	.94409	45
16	.26331	.96471	.28011	.95997	.29682	.95493	.31344	.94961	.32997	.94399	44
17	.26359	.96463	.28039	.95989	.29710	.95485	.31372	.94952	.33024	.94390	43
18	.26387	.96456	.28067	.95981	.29737	.95476	.31399	.94943	.33051	.94380	42
19	.26415	.96448	.28095	.95972	.29765	.95467	.31427	.94933	.33079	.94370	41
20	.26443	.96440	.28123	.95964	.29793	.95459	.31454	.94924	.33106	.94361	40
21	.26471	.96433	.28150	.95956	.29821	.95450	.31482	.94915	.33134	.94351	39
22	.26500	.96425	.28178	.95948	.29849	.95441	.31510	.94906	.33161	.94342	38
23	.26528	.96417	.28206	.95940	.29876	.95433	.31537	.94897	.33189	.94332	37
24	.26556	.96410	.28234	.95931	.29904	.95424	.31565	.94888	.33216	.94322	36
25	.26584	.96402	.28262	.95923	.29932	.95415	.31593	.94878	.33244	.94313	35
26	.26612	.96394	.28290	.95915	.29960	.95407	.31620	.94869	.33271	.94303	34
27	.26640	.96386	.28318	.95907	.29987	.95398	.31648	.94860	.33298	.94293	33
28	.26668	.96379	.28346	.95898	.30015	.95389	.31675	.94851	.33326	.94284	32
29	.26696	.96371	.28374	.95890	.30043	.95380	.31703	.94842	.33353	.94274	31
30	.26724	.96363	.28402	.95882	.30071	.95372	.31730	.94832	.33381	.94264	30
31	.26752	.96355	.28429	.95874	.30098	.95363	.31758	.94823	.33408	.94254	29
32	.26780	.96347	.28457	.95865	.30126	.95354	.31786	.94814	.33436	.94245	28
33	.26808	.96340	.28485	.95857	.30154	.95345	.31813	.94805	.33463	.94235	27
34	.26836	.96332	.28513	.95849	.30182	.95337	.31841	.94795	.33490	.94225	26
35	.26864	.96324	.28541	.95841	.30209	.95328	.31868	.94786	.33518	.94215	25
36	.26892	.96316	.28569	.95832	.30237	.95319	.31896	.94777	.33545	.94206	24
37	.26920	.96308	.28597	.95824	.30265	.95310	.31923	.94768	.33573	.94196	23
38	.26948	.96301	.28625	.95816	.30292	.95301	.31951	.94758	.33600	.94186	22
39	.26976	.96293	.28652	.95807	.30320	.95293	.31979	.94749	.33627	.94176	21
40	.27004	.96285	.28680	.95799	.30348	.95284	.32006	.94740	.33655	.94167	20
41	.27032	.96277	.28708	.95791	.30376	.95275	.32034	.94730	.33682	.94157	19
42	.27060	.96269	.28736	.95782	.30403	.95266	.32061	.94721	.33710	.94147	18
43	.27088	.96261	.28764	.95774	.30431	.95257	.32089	.94712	.33737	.94137	17
44	.27116	.96253	.28792	.95766	.30459	.95248	.32116	.94702	.33764	.94127	16
45	.27144	.96246	.28820	.95757	.30486	.95240	.32144	.94693	.33792	.94118	15
46	.27172	.96238	.28847	.95749	.30514	.95231	.32171	.94684	.33819	.94108	14
47	.27200	.96230	.28875	.95740	.30542	.95222	.32199	.94674	.33846	.94098	13
48	.27228	.96222	.28903	.95732	.30570	.95213	.32227	.94665	.33874	.94088	12
49	.27256	.96214	.28931	.95724	.30597	.95204	.32254	.94656	.33901	.94078	11
50	.27284	.96206	.28959	.95715	.30625	.95195	.32282	.94646	.33929	.94068	10
51	.27312	.96198	.28987	.95707	.30653	.95186	.32309	.94637	.33956	.94058	9
52	.27340	.96190	.29015	.95698	.30680	.95177	.32337	.94627	.33983	.94049	8
53	.27368	.96182	.29042	.95690	.30708	.95168	.32364	.94618	.34011	.94039	7
54	.27396	.96174	.29070	.95681	.30736	.95159	.32392	.94609	.34038	.94029	6
55	.27424	.96166	.29098	.95673	.30763	.95150	.32419	.94599	.34065	.94019	5
56	.27452	.96158	.29126	.95664	.30791	.95142	.32447	.94590	.34093	.94009	4
57	.27480	.96150	.29154	.95656	.30819	.95133	.32474	.93580	.34120	.93999	3
58	.27508	.96142	.29182	.95647	.30846	.95124	.32502	.94571	.34147	.93989	2
59	.27536	.96134	.29209	.95639	.30874	.95115	.32529	.94561	.34175	.93979	1
60	.27564	.96126	.29237	.95630	.30902	.95106	.32557	.94552	.34202	.93969	0
′	Cosin	Sine	Cosin	Sine	Cosin	Sine	Cosin	Sine	Cosin	Sine	′
	74°		73°		72°		71°		70°		

TABLE VI.—*Continued.*

NATURAL SINES AND COSINES.

′	20°		21°		22°		23°		24°		′
	Sine	Cosin	Sine	Cosin	Sine	Cosin	Sine	Cosin	Sine	Cosin	
0	.34202	.93969	.35837	.93358	.37461	.92718	.39073	.92050	.40674	.91355	60
1	.34229	.93959	.35864	.93348	.37488	.92707	.39100	.92039	.40700	.91343	59
2	.34257	.93949	.35891	.93337	.37515	.92697	.39127	.92028	.40727	.91331	58
3	.34284	.93939	.35918	.93327	.37542	.92686	.39153	.92016	.40753	.91319	57
4	.34311	.93929	.35945	.93316	.37569	.92675	.39180	.92005	.40780	.91307	56
5	.34339	.93919	.35973	.93306	.37595	.92664	.39207	.91994	.40806	.91295	55
6	.34366	.93909	.36000	.93295	.37622	.92653	.39234	.91982	.40833	.91283	54
7	.34393	.93899	.36027	.93285	.37649	.92642	.39260	.91971	.40860	.91272	53
8	.34421	.93889	.36054	.93274	.37676	.92631	.39287	.91959	.40886	.91260	52
9	.34448	.93879	.36081	.93264	.37703	.92620	.39314	.91948	.40913	.91248	51
10	.34475	.93869	.36108	.93253	.37730	.92609	.39341	.91936	.40939	.91236	50
11	.34503	.93859	.36135	.93243	.37757	.92598	.39367	.91925	.40966	.91224	49
12	.34530	.93849	.36162	.93232	.37784	.92587	.39394	.91914	.40992	.91212	48
13	.34557	.93839	.36190	.93222	.37811	.92576	.39421	.91902	.41019	.91200	47
14	.34584	.93829	.36217	.93211	.37838	.92565	.39448	.91891	.41045	.91188	46
15	.34612	.93819	.36244	.93201	.37865	.92554	.39474	.91879	.41072	.91176	45
16	.34639	.93809	.36271	.93190	.37892	.92543	.39501	.91868	.41098	.91164	44
17	.34666	.93799	.36298	.93180	.37919	.92532	.39528	.91856	.41125	.91152	43
18	.34694	.93789	.36325	.93169	.37946	.92521	.39555	.91845	.41151	.91140	42
19	.34721	.93779	.36352	.93159	.37973	.92510	.39581	.91833	.41178	.91128	41
20	.34748	.93769	.36379	.93148	.37999	.92499	.39608	.91822	.41204	.91116	40
21	.34775	.93759	.36406	.93137	.38026	.92488	.39635	.91810	.41231	.91104	39
22	.34803	.93748	.36434	.93127	.38053	.92477	.39661	.91799	.41257	.91092	38
23	.34830	.93738	.36461	.93116	.38080	.92466	.39688	.91787	.41284	.91080	37
24	.34857	.93728	.36488	.93106	.38107	.92455	.39715	.91775	.41310	.91068	36
25	.34884	.93718	.36515	.93095	.38134	.92444	.39741	.91764	.41337	.91056	35
26	.34912	.93708	.36542	.93084	.38161	.92432	.39768	.91752	.41363	.91044	34
27	.34939	.93698	.36569	.93074	.38188	.92421	.39795	.91741	.41390	.91032	33
28	.34966	.93688	.36596	.93063	.38215	.92410	.39822	.91729	.41416	.91020	32
29	.34993	.93677	.36623	.93052	.38241	.92399	.39848	.91718	.41443	.91008	31
30	.35021	.93667	.36650	.93042	.38268	.92388	.39875	.91706	.41469	.90996	30
31	.35048	.93657	.36677	.93031	.38295	.92377	.39902	.91694	.41496	.90984	29
32	.35075	.93647	.36704	.93020	.38322	.92366	.39928	.91683	.41522	.90972	28
33	.35102	.93637	.36731	.93010	.38349	.92355	.39955	.91671	.41549	.90960	27
34	.35130	.93626	.36758	.92999	.38376	.92343	.39982	.91660	.41575	.90948	26
35	.35157	.93616	.36785	.92988	.38403	.92332	.40008	.91648	.41602	.90936	25
36	.35184	.93606	.36812	.92978	.38430	.92321	.40035	.91636	.41628	.90924	24
37	.35211	.93596	.36839	.92967	.38456	.92310	.40062	.91625	.41655	.90911	23
38	.35239	.93585	.36867	.92956	.38483	.92299	.40088	.91613	.41681	.90899	22
39	.35266	.93575	.36894	.92945	.38510	.92287	.40115	.91601	.41707	.90887	21
40	.35293	.93565	.36921	.92935	.38537	.92276	.40141	.91590	.41734	.90875	20
41	.35320	.93555	.36948	.92924	.38564	.92265	.40168	.91578	.41760	.90863	19
42	.35347	.93544	.36975	.92913	.38591	.92254	.40195	.91566	.41787	.90851	18
43	.35375	.93534	.37002	.92902	.38617	.92243	.40221	.91555	.41813	.90839	17
44	.35402	.93524	.37029	.92892	.38644	.92231	.40248	.91543	.41840	.90826	16
45	.35429	.93514	.37056	.92881	.38671	.92220	.40275	.91531	.41866	.90814	15
46	.35456	.93503	.37083	.92870	.38698	.92209	.40301	.91519	.41892	.90802	14
47	.35484	.93493	.37110	.92859	.38725	.92198	.40328	.91508	.41919	.90790	13
48	.35511	.93483	.37137	.92849	.38752	.92186	.40355	.91496	.41945	.90778	12
49	.35538	.93472	.37164	.92838	.38778	.92175	.40381	.91484	.41972	.90766	11
50	.35565	.93462	.37191	.92827	.38805	.92164	.40408	.91472	.41998	.90753	10
51	.35592	.93452	.37218	.92816	.38832	.92152	.40434	.91461	.42024	.90741	9
52	.35619	.93441	.37245	.92805	.38859	.92141	.40461	.91449	.42051	.90729	8
53	.35647	.93431	.37272	.92794	.38886	.92130	.40488	.91437	.42077	.90717	7
54	.35674	.93420	.37299	.92784	.38912	.92119	.40514	.91425	.42104	.90704	6
55	.35701	.93410	.37326	.92773	.38939	.92107	.40541	.91414	.42130	.90692	5
56	.35728	.93400	.37353	.92762	.38966	.92096	.40567	.91402	.42156	.90680	4
57	.35755	.93389	.37380	.92751	.38993	.92085	.40594	.91390	.42183	.90668	3
58	.35782	.93379	.37407	.92740	.39020	.92073	.40621	.91378	.42209	.90655	2
59	.35810	.93368	.37434	.92729	.39046	.92062	.40647	.91366	.42235	.90643	1
60	.35837	.93358	.37461	.92718	.39073	.92050	.40674	.91355	.42262	.90631	0
	Cosin	Sine	Cosin	Sine	Cosin	Sine	Cosin	Sine	Cosin	Sine	
′	69°		68°		67°		66°		65°		′

TABLE VI.—*Continued.*

NATURAL SINES AND COSINES.

′	25°		26°		27°		28°		29°		′
	Sine	Cosin	Sine	Cosin	Sine	Cosin	Sine	Cosin	Sine	Cosin	
0	.42262	.90631	.43837	.89879	.45399	.89101	.46947	.88295	.48481	.87462	60
1	.42288	.90618	.43863	.89867	.45425	.89087	.46973	.88281	.48506	.87448	59
2	.42315	.90606	.43889	.89854	.45451	.89074	.46999	.88267	.48532	.87434	58
3	.42341	.90594	.43916	.89841	.45477	.89061	.47024	.88254	.48557	.87420	57
4	.42367	.90582	.43942	.89828	.45503	.89048	.47050	.88240	.48583	.87406	56
5	.42394	.90569	.43968	.89816	.45529	.89035	.47076	.88226	.48608	.87391	55
6	.42420	.90557	.43994	.89803	.45554	.89021	.47101	.88213	.48634	.87377	54
7	.42446	.90545	.44020	.89790	.45580	.89008	.47127	.88199	.48659	.87363	53
8	.42473	.90532	.44046	.89777	.45606	.88995	.47153	.88185	.48684	.87349	52
9	.42499	.90520	.44072	.89764	.45632	.88981	.47178	.88172	.48710	.87335	51
10	.42525	.90507	.44098	.89752	.45658	.88968	.47204	.88158	.48735	.87321	50
11	.42552	.90495	.44124	.89739	.45684	.88955	.47229	.88144	.48761	.87306	49
12	.42578	.90483	.44151	.89726	.45710	.88942	.47255	.88130	.48786	.87292	48
13	.42604	.90470	.44177	.89713	.45736	.88928	.47281	.88117	.48811	.87278	47
14	.42631	.90458	.44203	.89700	.45762	.88915	.47306	.88103	.48837	.87264	46
15	.42657	.90446	.44229	.89687	.45787	.88902	.47332	.88089	.48862	.87250	45
16	.42683	.90433	.44255	.89674	.45813	.88888	.47358	.88075	.48888	.87235	44
17	.42709	.90421	.44281	.89662	.45839	.88875	.47383	.88062	.48913	.87221	43
18	.42736	.90408	.44307	.89649	.45865	.88862	.47409	.88048	.48938	.87207	42
19	.42762	.90396	.44333	.89636	.45891	.88848	.47434	.88034	.48964	.87193	41
20	.42788	.90383	.44359	.89623	.45917	.88835	.47460	.88020	.48989	.87178	40
21	.42815	.90371	.44385	.89610	.45942	.88822	.47486	.88006	.49014	.87164	39
22	.42841	.90358	.44411	.89597	.45968	.88808	.47511	.87993	.49040	.87150	38
23	.42867	.90346	.44437	.89584	.45994	.88795	.47537	.87979	.49065	.87136	37
24	.42894	.90334	.44464	.89571	.46020	.88782	.47562	.87965	.49090	.87121	36
25	.42920	.90321	.44490	.89558	.46046	.88768	.47588	.87951	.49116	.87107	35
26	.42946	.90309	.44516	.89545	.46072	.88755	.47614	.87937	.49141	.87093	34
27	.42972	.90296	.44542	.89532	.46097	.88741	.47639	.87923	.49166	.87079	33
28	.42999	.90284	.44568	.89519	.46123	.88728	.47665	.87909	.49192	.87064	32
29	.43025	.90271	.44594	.89506	.46149	.88715	.47690	.87896	.49217	.87050	31
30	.43051	.90259	.44620	.89493	.46175	.88701	.47716	.87882	.49242	.87036	30
31	.43077	.90246	.44646	.89480	.46201	.88688	.47741	.87868	.49268	.87021	29
32	.43104	.90233	.44672	.89467	.46226	.88674	.47767	.87854	.49293	.87007	28
33	.43130	.90221	.44698	.89454	.46252	.88661	.47793	.87840	.49318	.86993	27
34	.43156	.90208	.44724	.89441	.46278	.88647	.47818	.87826	.49344	.86978	26
35	.43182	.90196	.44750	.89428	.46304	.88634	.47844	.87812	.49369	.86964	25
36	.43209	.90183	.44776	.89415	.46330	.88620	.47869	.87798	.49394	.86949	24
37	.43235	.90171	.44802	.89402	.46355	.88607	.47895	.87784	.49419	.86935	23
38	.43261	.90158	.44828	.89389	.46381	.88593	.47920	.87770	.49445	.86921	22
39	.43287	.90146	.44854	.89376	.46407	.88580	.47946	.87756	.49470	.86906	21
40	.43313	.90133	.44880	.89363	.46433	.88566	.47971	.87743	.49495	.86892	20
41	.43340	.90120	.44906	.89350	.46458	.88553	.47997	.87729	.49521	.86878	19
42	.43366	.90108	.44932	.89337	.46484	.88539	.48022	.87715	.49546	.86863	18
43	.43392	.90095	.44958	.89324	.46510	.88526	.48048	.87701	.49571	.86849	17
44	.43418	.90082	.44984	.89311	.46536	.88512	.48073	.87687	.49596	.86834	16
45	.43445	.90070	.45010	.89298	.46561	.88499	.48099	.87673	.49622	.86820	15
46	.43471	.90057	.45036	.89285	.46587	.88485	.48124	.87659	.49647	.86805	14
47	.43497	.90045	.45062	.89272	.46613	.88472	.48150	.87645	.49672	.86791	13
48	.43523	.90032	.45088	.89259	.46639	.88458	.48175	.87631	.49697	.86777	12
49	.43549	.90019	.45114	.89245	.46664	.88445	.48201	.87617	.49723	.86762	11
50	.43575	.90007	.45140	.89232	.46690	.88431	.48226	.87603	.49748	.86748	10
51	.43602	.89994	.45166	.89219	.46716	.88417	.48252	.87589	.49773	.86733	9
52	.43628	.89981	.45192	.89206	.46742	.88404	.48277	.87575	.49798	.86719	8
53	.43654	.89968	.45218	.89193	.46767	.88390	.48303	.87561	.49824	.86704	7
54	.43680	.89956	.45243	.89180	.46793	.88377	.48328	.87546	.49849	.86690	6
55	.43706	.89943	.45269	.89167	.46819	.88363	.48354	.87532	.49874	.86675	5
56	.43733	.89930	.45295	.89153	.46844	.88349	.48379	.87518	.49899	.86661	4
57	.43759	.89918	.45321	.89140	.46870	.88336	.48405	.87504	.49924	.86646	3
58	.43785	.89905	.45347	.89127	.46896	.88322	.48430	.87490	.49950	.86632	2
59	.43811	.89892	.45373	.89114	.46921	.88308	.48456	.87476	.49975	.86617	1
60	.43837	.89879	.45399	.89101	.46947	.88295	.48481	.87462	.50000	.86603	0
′	Cosin	Sine	Cosin	Sine	Cosin	Sine	Cosin	Sine	Cosin	Sine	′
	64°		63°		62°		61°		60°		

TABLE VI.—*Continued.*

NATURAL SINES AND COSINES.

′	30° Sine	30° Cosin	31° Sine	31° Cosin	32° Sine	32° Cosin	33° Sine	33° Cosin	34° Sine	34° Cosin	′
0	.50000	.86603	.51504	.85717	.52992	.84805	.54464	.83867	.55919	.82904	60
1	.50025	.86588	.51529	.85702	.53017	.84789	.54488	.83851	.55943	.82887	59
2	.50050	.86573	.51554	.85687	.53041	.84774	.54513	.83835	.55968	.82871	58
3	.50076	.86559	.51579	.85672	.53066	.84759	.54537	.83819	.55992	.82855	57
4	.50101	.86544	.51604	.85657	.53091	.84743	.54561	.83804	.56016	.82839	56
5	.50126	.86530	.51628	.85642	.53115	.84728	.54586	.83788	.56040	.82822	55
6	.50151	.86515	.51653	.85627	.53140	.84712	.54610	.83772	.56064	.82806	54
7	.50176	.86501	.51678	.85612	.53164	.84697	.54635	.83756	.56088	.82790	53
8	.50201	.86486	.51703	.85597	.53189	.84681	.54659	.83740	.56112	.82773	52
9	.50227	.86471	.51728	.85582	.53214	.84666	.54683	.83724	.56136	.82757	51
10	.50252	.86457	.51753	.85567	.53238	.84650	.54708	.83708	.56160	.82741	50
11	.50277	.86442	.51778	.85551	.53263	.84635	.54732	.83692	.56184	.82724	49
12	.50302	.86427	.51803	.85536	.53288	.84619	.54756	.83676	.56208	.82708	48
13	.50327	.86413	.51828	.85521	.53312	.84604	.54781	.83660	.56232	.82692	47
14	.50352	.86398	.51852	.85506	.53337	.84588	.54805	.83645	.56256	.82675	46
15	.50377	.86384	.51877	.85491	.53361	.84573	.54829	.83629	.56280	.82659	45
16	.50403	.86369	.51902	.85476	.53386	.84557	.54854	.83613	.56305	.82643	44
17	.50428	.86354	.51927	.85461	.53411	.84542	.54878	.83597	.56329	.82626	43
18	.50453	.86340	.51952	.85446	.53435	.84526	.54902	.83581	.56353	.82610	42
19	.50478	.86325	.51977	.85431	.53460	.84511	.54927	.83565	.56377	.82593	41
20	.50503	.86310	.52002	.85416	.53484	.84495	.54951	.83549	.56401	.82577	40
21	.50528	.86295	.52026	.85401	.53509	.84480	.54975	.83533	.56425	.82561	39
22	.50553	.86281	.52051	.85385	.53534	.84464	.54999	.83517	.56449	.82544	38
23	.50578	.86266	.52076	.85370	.53558	.84448	.55024	.83501	.56473	.82528	37
24	.50603	.86251	.52101	.85355	.53583	.84433	.55048	.83485	.56497	.82511	36
25	.50628	.86237	.52126	.85340	.53607	.84417	.55072	.83469	.56521	.82495	35
26	.50654	.86222	.52151	.85325	.53632	.84402	.55097	.83453	.56545	.82478	34
27	.50679	.86207	.52175	.85310	.53656	.84386	.55121	.83437	.56569	.82462	33
28	.50704	.86192	.52200	.85294	.53681	.84370	.55145	.83421	.56593	.82446	32
29	.50729	.86178	.52225	.85279	.53705	.84355	.55169	.83405	.56617	.82429	31
30	.50754	.86163	.52250	.85264	.53730	.84339	.55194	.83389	.56641	.82413	30
31	.50779	.86148	.52275	.85249	.53754	.84324	.55218	.83373	.56665	.82396	29
32	.50804	.86133	.52299	.85234	.53779	.84308	.55242	.83356	.56689	.82380	28
33	.50829	.86119	.52324	.85218	.53804	.84292	.55266	.83340	.56713	.82363	27
34	.50854	.86104	.52349	.85203	.53828	.84277	.55291	.83324	.56736	.82347	26
35	.50879	.86089	.52374	.85188	.53853	.84261	.55315	.83308	.56760	.82330	25
36	.50904	.86074	.52399	.85173	.53877	.84245	.55339	.83292	.56784	.82314	24
37	.50929	.86059	.52423	.85157	.53902	.84230	.55363	.83276	.56808	.82297	23
38	.50954	.86045	.52448	.85142	.53926	.84214	.55388	.83260	.56832	.82281	22
39	.50979	.86030	.52473	.85127	.53951	.84198	.55412	.83244	.56856	.82264	21
40	.51004	.86015	.52498	.85112	.53975	.84182	.55436	.83228	.56880	.82248	20
41	.51029	.86000	.52522	.85096	.54000	.84167	.55460	.83212	.56904	.82231	19
42	.51054	.85985	.52547	.85081	.54024	.84151	.55484	.83195	.56928	.82214	18
43	.51079	.85970	.52572	.85066	.54049	.84135	.55509	.83179	.56952	.82198	17
44	.51104	.85956	.52597	.85051	.54073	.84120	.55533	.83163	.56976	.82181	16
45	.51129	.85941	.52621	.85035	.54097	.84104	.55557	.83147	.57000	.82165	15
46	.51154	.85926	.52646	.85020	.54122	.84088	.55581	.83131	.57024	.82148	14
47	.51179	.85911	.52671	.85005	.54146	.84072	.55605	.83115	.57047	.82132	13
48	.51204	.85896	.52696	.84989	.54171	.84057	.55630	.83098	.57071	.82115	12
49	.51229	.85881	.52720	.84974	.54195	.84041	.55654	.83082	.57095	.82098	11
50	.51254	.85866	.52745	.84959	.54220	.84025	.55678	.83066	.57119	.82082	10
51	.51279	.85851	.52770	.84943	.54244	.84009	.55702	.83050	.57143	.82065	9
52	.51304	.85836	.52794	.84928	.54269	.83994	.55726	.83034	.57167	.82048	8
53	.51329	.85821	.52819	.84913	.54293	.83978	.55750	.83017	.57191	.82032	7
54	.51354	.85806	.52844	.84897	.54317	.83962	.55775	.83001	.57215	.82015	6
55	.51379	.85792	.52869	.84882	.54342	.83946	.55799	.82985	.57238	.81999	5
56	.51404	.85777	.52893	.84866	.54366	.83930	.55823	.82969	.57262	.81982	4
57	.51429	.85762	.52918	.84851	.54391	.83915	.55847	.82953	.57286	.81965	3
58	.51454	.85747	.52943	.84836	.54415	.83899	.55871	.82936	.57310	.81949	2
59	.51479	.85732	.52967	.84820	.54440	.83883	.55895	.82920	.57334	.81932	1
60	.51504	.85717	.52992	.84805	.54464	.83867	.55919	.82904	.57358	.81915	0
′	Cosin	Sine	Cosin	Sine	Cosin	Sine	Cosin	Sine	Cosin	Sine	′
	59°		58°		57°		56°		55°		

TABLE VI.—*Continued.*

Natural Sines and Cosines.

′	35°		36°		37°		38°		39°		′
	Sine	Cosin	Sine	Cosin	Sine	Cosin	Sine	Cosin	Sine	Cosin	
0	.57358	.81915	.58779	.80902	.60182	.79864	.61566	.78801	.62932	.77715	60
1	.57381	.81899	.58802	.80885	.60205	.79846	.61589	.78783	.62955	.77696	59
2	.57405	.81882	.58826	.80867	.60228	.79829	.61612	.78765	.62977	.77678	58
3	.57429	.81865	.58849	.80850	.60251	.79811	.61635	.78747	.63000	.77660	57
4	.57453	.81848	.58873	.80833	.60274	.79793	.61658	.78729	.63022	.77641	56
5	.57477	.81832	.58896	.80816	.60298	.79776	.61681	.78711	.63045	.77623	55
6	.57501	.81815	.58920	.80799	.60321	.79758	.61704	.78694	.63068	.77605	54
7	.57524	.81798	.58943	.80782	.60344	.79741	.61726	.78676	.63090	.77586	53
8	.57548	.81782	.58967	.80765	.60367	.79723	.61749	.78658	.63113	.77568	52
9	.57572	.81765	.58990	.80748	.60390	.79706	.61772	.78640	.63135	.77550	51
10	.57596	.81748	.59014	.80730	.60414	.79688	.61795	.78622	.63158	.77531	50
11	.57619	.81731	.59037	.80713	.60437	.79671	.61818	.78604	.63180	.77513	40
12	.57643	.81714	.59061	.80696	.60460	.79653	.61841	.78586	.63203	.77494	48
13	.57667	.81698	.59084	.80679	.60483	.79635	.61864	.78568	.63225	.77476	47
14	.57691	.81681	.59108	.80662	.60506	.79618	.61887	.78550	.63248	.77458	46
15	.57715	.81664	.59131	.80644	.60529	.79600	.61909	.78532	.63271	.77439	45
16	.57738	.81647	.59154	.80627	.60553	.79583	.61932	.78514	.63293	.77421	44
17	.57762	.81631	.59178	.80610	.60576	.79565	.61955	.78496	.63316	.77402	43
18	.57786	.81614	.59201	.80593	.60599	.79547	.61978	.78478	.63338	.77384	42
19	.57810	.81597	.59225	.80576	.60622	.79530	.62001	.78460	.63361	.77366	41
20	.57833	.81580	.59248	.80558	.60645	.79512	.62024	.78442	.63383	.77347	40
21	.57857	.81563	.59272	.80541	.60668	.79494	.62046	.78424	.63406	.77329	39
22	.57881	.81546	.59295	.80524	.60691	.79477	.62069	.78405	.63428	.77310	38
23	.57904	.81530	.59318	.80507	.60714	.79459	.62092	.78387	.63451	.77292	37
24	.57928	.81513	.59342	.80489	.60738	.79441	.62115	.78369	.63473	.77273	36
25	.57952	.81496	.59365	.80472	.60761	.79424	.62138	.78351	.63496	.77255	35
26	.57976	.81479	.59389	.80455	.60784	.79406	.62160	.78333	.63518	.77236	34
27	.57999	.81462	.59412	.80438	.60807	.79388	.62183	.78315	.63540	.77218	33
28	.58023	.81445	.59436	.80420	.60830	.79371	.62206	.78297	.63563	.77199	32
29	.58047	.81428	.59459	.80403	.60853	.79353	.62229	.78279	.63585	.77181	31
30	.58070	.81412	.59482	.80386	.60876	.79335	.62251	.78261	.63608	.77162	30
31	.58094	.81395	.59506	.80368	.60899	.79318	.62274	.78243	.63630	.77144	29
32	.58118	.81378	.59529	.80351	.60922	.79300	.62297	.78225	.63653	.77125	28
33	.58141	.81361	.59552	.80334	.60945	.79282	.62320	.78206	.63675	.77107	27
34	.58165	.81344	.59576	.80316	.60968	.79264	.62342	.78188	.63698	.77088	26
35	.58189	.81327	.59599	.80299	.60991	.79247	.62365	.78170	.63720	.77070	25
36	.58212	.81310	.59622	.80282	.61015	.79229	.62388	.78152	.63742	.77051	24
37	.58236	.81293	.59646	.80264	.61038	.79211	.62411	.78134	.63765	.77033	23
38	.58260	.81276	.59669	.80247	.61061	.79193	.62433	.78116	.63787	.77014	22
39	.58283	.81259	.59693	.80230	.61084	.79176	.62456	.78098	.63810	.76996	21
40	.58307	.81242	.59716	.80212	.61107	.79158	.62479	.78079	.63832	.76977	20
41	.58330	.81225	.59739	.80195	.61130	.79140	.62502	.78061	.63854	.76959	19
42	.58354	.81208	.59763	.80178	.61153	.79122	.62524	.78043	.63877	.76940	18
43	.58378	.81191	.59786	.80160	.61176	.79105	.62547	.78025	.63899	.76921	17
44	.58401	.81174	.59809	.80143	.61199	.79087	.62570	.78007	.63922	.76903	16
45	.58425	.81157	.59832	.80125	.61222	.79069	.62592	.77988	.63944	.76884	15
46	.58449	.81140	.59856	.80108	.61245	.79051	.62615	.77970	.63966	.76866	14
47	.58472	.81123	.59879	.80091	.61268	.79033	.62638	.77952	.63989	.76847	13
48	.58496	.81106	.59902	.80073	.61291	.79016	.62660	.77934	.64011	.76828	12
49	.58519	.81089	.59926	.80056	.61314	.78998	.62683	.77916	.64033	.76810	11
50	.58543	.81072	.59949	.80038	.61337	.78980	.62706	.77897	.64056	.76791	10
51	.58567	.81055	.59972	.80021	.61360	.78962	.62728	.77879	.64078	.76772	9
52	.58590	.81038	.59995	.80003	.61383	.78944	.62751	.77861	.64100	.76754	8
53	.58614	.81021	.60019	.79986	.61406	.78926	.62774	.77843	.64123	.76735	7
54	.58637	.81004	.60042	.79968	.61429	.78908	.62796	.77824	.64145	.76717	6
55	.58661	.80987	.60065	.79951	.61451	.78891	.62819	.77806	.64167	.76698	5
56	.58684	.80970	.60089	.79934	.61474	.78873	.62842	.77788	.64190	.76679	4
57	.58708	.80953	.60112	.79916	.61497	.78855	.62864	.77769	.64212	.76661	3
58	.58731	.80936	.60135	.79899	.61520	.78837	.62887	.77751	.64234	.76642	2
59	.58755	.80919	.60158	.79881	.61543	.78819	.62909	.77733	.64256	.76623	1
60	.58779	.80902	.60182	.79864	.61566	.78801	.62932	.77715	.64279	.76604	0
′	Cosin	Sine	Cosin	Sine	Cosin	Sine	Cosin	Sine	Cosin	Sine	′
	54°		53°		52°		51°		50°		

TABLE VI.—*Continued.*

NATURAL SINES AND COSINES.

′	40° Sine	40° Cosin	41° Sine	41° Cosin	42° Sine	42° Cosin	43° Sine	43° Cosin	44° Sine	44° Cosin	′
0	.64279	.76604	.65606	.75471	.66913	.74314	.68200	.73135	.69466	.71934	60
1	.64301	.76586	.65628	.75452	.66935	.74295	.68221	.73116	.69487	.71914	59
2	.64323	.76567	.65650	.75433	.66956	.74276	.68242	.73096	.69508	.71894	58
3	.64346	.76548	.65672	.75414	.66978	.74256	.68264	.73076	.69529	.71873	57
4	.64368	.76530	.65694	.75395	.66999	.74237	.68285	.73056	.69549	.71853	56
5	.64390	.76511	.65716	.75375	.67021	.74217	.68306	.73036	.69570	.71833	55
6	.64412	.76492	.65738	.75356	.67043	.74198	.68327	.73016	.69591	.71813	54
7	.64435	.76473	.65759	.75337	.67064	.74178	.68349	.72996	.69612	.71792	53
8	.64457	.76455	.65781	.75318	.67086	.74159	.68370	.72976	.69633	.71772	52
9	.64479	.76436	.65803	.75299	.67107	.74139	.68391	.72957	.69654	.71752	51
10	.64501	.76417	.65825	.75280	.67129	.74120	.68412	.72937	.69675	.71732	50
11	.64524	.76398	.65847	.75261	.67151	.74100	.68434	.72917	.69696	.71711	49
12	.64546	.76380	.65869	.75241	.67172	.74080	.68455	.72897	.69717	.71691	48
13	.64568	.76361	.65891	.75222	.67194	.74061	.68476	.72877	.69737	.71671	47
14	.64590	.76342	.65913	.75203	.67215	.74041	.68497	.72857	.69758	.71650	46
15	.64612	.76323	.65935	.75184	.67237	.74022	.68518	.72837	.69779	.71630	45
16	.64635	.76304	.65956	.75165	.67258	.74002	.68539	.72817	.69800	.71610	44
17	.64657	.76286	.65978	.75146	.67280	.73983	.68561	.72797	.69821	.71590	43
18	.64679	.76267	.66000	.75126	.67301	.73963	.68582	.72777	.69842	.71569	42
19	.64701	.76248	.66022	.75107	.67323	.73944	.68603	.72757	.69862	.71549	41
20	.64723	.76229	.66044	.75088	.67344	.73924	.68624	.72737	.69883	.71529	40
21	.64746	.76210	.66066	.75069	.67366	.73904	.68645	.72717	.69904	.71508	39
22	.64768	.76192	.66088	.75050	.67387	.73885	.68666	.72697	.69925	.71488	38
23	.64790	.76173	.66109	.75030	.67409	.73865	.68688	.72677	.69946	.71468	37
24	.64812	.76154	.66131	.75011	.67430	.73846	.68709	.72657	.69966	.71447	36
25	.64834	.76135	.66153	.74992	.67452	.73826	.68730	.72637	.69987	.71427	35
26	.64856	.76116	.66175	.74973	.67473	.73806	.68751	.72617	.70008	.71407	34
27	.64878	.76097	.66197	.74953	.67495	.73787	.68772	.72597	.70029	.71386	33
28	.64901	.76078	.66218	.74934	.67516	.73767	.68793	.72577	.70049	.71366	32
29	.64923	.76059	.66240	.74915	.67538	.73747	.68814	.72557	.70070	.71345	31
30	.64945	.76041	.66262	.74896	.67559	.73728	.68835	.72537	.70091	.71325	30
31	.64967	.76022	.66284	.74876	.67580	.73708	.68857	.72517	.70112	.71305	29
32	.64989	.76003	.66306	.74857	.67602	.73688	.68878	.72497	.70132	.71284	28
33	.65011	.75984	.66327	.74838	.67623	.73669	.68899	.72477	.70153	.71264	27
34	.65033	.75965	.66349	.74818	.67645	.73649	.68920	.72457	.70174	.71243	26
35	.65055	.75946	.66371	.74799	.67666	.73629	.68941	.72437	.70195	.71223	25
36	.65077	.75927	.66393	.74780	.67688	.73610	.68962	.72417	.70215	.71203	24
37	.65100	.75908	.66414	.74760	.67709	.73590	.68983	.72397	.70236	.71182	23
38	.65122	.75889	.66436	.74741	.67730	.73570	.69004	.72377	.70257	.71162	22
39	.65144	.75870	.66458	.74722	.67752	.73551	.69025	.72357	.70277	.71141	21
40	.65166	.75851	.66480	.74703	.67773	.73531	.69046	.72337	.70298	.71121	20
41	.65188	.75832	.66501	.74683	.67795	.73511	.69067	.72317	.70319	.71100	19
42	.65210	.75813	.66523	.74664	.67816	.73491	.69088	.72297	.70339	.71080	18
43	.65232	.75794	.66545	.74644	.67837	.73472	.69109	.72277	.70360	.71059	17
44	.65254	.75775	.66566	.74625	.67859	.73452	.69130	.72257	.70381	.71039	16
45	.65276	.75756	.66588	.74606	.67880	.73432	.69151	.72236	.70401	.71019	15
46	.65298	.75738	.66610	.74586	.67901	.73413	.69172	.72216	.70422	.70998	14
47	.65320	.75719	.66632	.74567	.67923	.73393	.69193	.72196	.70443	.70978	13
48	.65342	.75700	.66653	.74548	.67944	.73373	.69214	.72176	.70463	.70957	12
49	.65364	.75680	.66675	.74528	.67965	.73353	.69235	.72156	.70484	.70937	11
50	.65386	.75661	.66697	.74509	.67987	.73333	.69256	.72136	.70505	.70916	10
51	.65408	.75642	.66718	.74489	.68008	.73314	.69277	.72116	.70525	.70896	9
52	.65430	.75623	.66740	.74470	.68029	.73294	.69298	.72095	.70546	.70875	8
53	.65452	.75604	.66762	.74451	.68051	.73274	.69319	.72075	.70567	.70855	7
54	.65474	.75585	.66783	.74431	.68072	.73254	.69340	.72055	.70587	.70834	6
55	.65496	.75566	.66805	.74412	.68093	.73234	.69361	.72035	.70608	.70813	5
56	.65518	.75547	.66827	.74392	.68115	.73215	.69382	.72015	.70628	.70793	4
57	.65540	.75528	.66848	.74373	.68136	.73195	.69403	.71995	.70649	.70772	3
58	.65562	.75509	.66870	.74352	.68157	.73175	.69424	.71974	.70670	.70752	2
59	.65584	.75490	.66891	.74334	.68179	.73155	.69445	.71954	.70690	.70731	1
60	.65606	.75471	.66913	.74314	.68200	.73135	.69466	.71934	.70711	.70711	0
′	Cosin 49°	Sine 49°	Cosin 48°	Sine 48°	Cosin 47°	Sine 47°	Cosin 46°	Sine 46°	Cosin 45°	Sine 45°	′

TABLE VII.

NATURAL TANGENTS AND COTANGENTS.

′	0°		1°		2°		3°		′
	Tang	Cotang	Tang	Cotang	Tang	Cotang	Tang	Cotang	
0	.00000	Infinite.	.01746	57.2900	.03492	28.6363	.05241	19.0811	60
1	.00029	3437.75	.01775	56.3506	.03521	28.3994	.05270	18.9755	59
2	.00058	1718.87	.01804	55.4415	.03550	28.1664	.05299	18.8711	58
3	.00087	1145.92	.01833	54.5613	.03579	27.9372	.05328	18.7678	57
4	.00116	859.436	.01862	53.7086	.03609	27.7117	.05357	18.6656	56
5	.00145	687.549	.01891	52.8821	.03638	27.4899	.05387	18.5645	55
6	.00175	572.957	.01920	52.0807	.03667	27.2715	.05416	18.4645	54
7	.00204	491.106	.01949	51.3032	.03696	27.0566	.05445	18.3655	53
8	.00233	429.718	.01978	50.5485	.03725	26.8450	.05474	18.2677	52
9	.00262	381.971	.02007	49.8157	.03754	26.6367	.05503	18.1708	51
10	.00291	343.774	.02036	49.1039	.03783	26.4316	.05533	18.0750	50
11	.00320	312.521	.02066	48.4121	.03812	26.2296	.05562	17.9802	49
12	.00349	286.478	.02095	47.7395	.03842	26.0307	.05591	17.8863	48
13	.00378	204.441	.02124	47.0853	.03871	25.8348	.05620	17.7934	47
14	.00407	245.552	.02153	46.4489	.03900	25.6418	.05649	17.7015	46
15	.00436	229.182	.02182	45.8294	.03929	25.4517	.05678	17.6106	45
16	.00465	214.858	.02211	45.2261	.03958	25.2644	.05708	17.5205	44
17	.00495	202.219	.02240	44.6386	.03987	25.0798	.05737	17.4314	43
18	.00524	190.984	.02269	44.0661	.04016	24.8978	.05766	17.3432	42
19	.00553	180.932	.02298	43.5081	.04046	24.7185	.05795	17.2558	41
20	.00582	171.885	.02328	42.9641	.04075	24.5418	.05824	17.1693	40
21	.00611	163.700	.02357	42.4335	.04104	24.3675	.05854	17.0837	39
22	.00640	156.259	.02386	41.9158	.04133	24.1957	.05883	16.9990	38
23	.00669	149.465	.02415	41.4106	.04162	24.0263	.05912	16.9150	37
24	.00698	143.237	.02444	40.9174	.04191	23.8593	.05941	16.8319	36
25	.00727	137.507	.02473	40.4358	.04220	23.6945	.05970	16.7496	35
26	.00756	132.219	.02502	39.9655	.04250	23.5321	.05999	16.6681	34
27	.00785	127.321	.02531	39.5059	.04279	23.3718	.06029	16.5874	33
28	.00815	122.774	.02560	39.0568	.04308	23.2137	.06058	16.5075	32
29	.00844	118.540	.02589	38.6177	.04337	23.0577	.06087	16.4283	31
30	.00873	114.589	.02619	38.1885	.04366	22.9038	.06116	16.3499	30
31	.00902	110.892	.02648	37.7686	.04395	22.7519	.06145	16.2722	29
32	.00931	107.426	.02677	37.3579	.04424	22.6020	.06175	16.1952	28
33	.00960	104.171	.02706	36.9560	.04454	22.4541	.06204	16.1190	27
34	.00989	101.107	.02735	36.5627	.04483	22.3081	.06233	16.0435	26
35	.01018	98.2179	.02764	36.1776	.04512	22.1640	.06262	15.9687	25
36	.01047	95.4895	.02793	35.8006	.04541	22.0217	.06291	15.8945	24
37	.01076	92.9085	.02822	35.4313	.04570	21.8813	.06321	15.8211	23
38	.01105	90.4633	.02851	35.0695	.04599	21.7426	.06350	15.7483	22
39	.01135	88.1436	.02881	34.7151	.04628	21.6056	.06379	15.6762	21
40	.01164	85.9398	.02910	34.3678	.04658	21.4704	.06408	15.6048	20
41	.01193	83.8435	.02939	34.0273	.04687	21.3369	.06437	15.5340	19
42	.01222	81.8470	.02968	33.6935	.04716	21.2049	.06467	15.4638	18
43	.01251	79.9434	.02997	33.3662	.04745	21.0747	.06496	15.3943	17
44	.01280	78.1263	.03026	33.0452	.04774	20.9460	.06525	15.3254	16
45	.01309	76.3900	.03055	32.7303	.04803	20.8188	.06554	15.2571	15
46	.01338	74.7292	.03084	32.4213	.04833	20.6932	.06584	15.1893	14
47	.01367	73.1390	.03114	32.1181	.04862	20.5691	.06613	15.1222	13
48	.01396	71.6151	.03143	31.8205	.04891	20.4465	.06642	15.0557	12
49	.01425	70.1533	.03172	31.5284	.04920	20.3253	.06671	14.9898	11
50	.01455	68.7501	.03201	31.2416	.04949	20.2056	.06700	14.9244	10
51	.01484	67.4019	.03230	30.9599	.04978	20.0872	.06730	14.8596	9
52	.01513	66.1055	.03259	30.6833	.05007	19.9702	.06759	14.7954	8
53	.01542	64.8580	.03288	30.4116	.05037	19.8546	.06788	14.7317	7
54	.01571	63.6567	.03317	30.1446	.05066	19.7403	.06817	14.6685	6
55	.01600	62.4992	.03346	29.8823	.05095	19.6273	.06847	14.6059	5
56	.01629	61.3829	.03376	29.6245	.05124	19.5156	.06876	14.5438	4
57	.01658	60.3058	.03405	29.3711	.05153	19.4051	.06905	14.4823	3
58	.01687	59.2659	.03434	29.1220	.05182	19.2959	.06934	14.4212	2
59	.01716	58.2612	.03463	28.8771	.05212	19.1879	.06963	14.3607	1
60	.01746	57.2900	.03492	28.6363	.05241	19.0811	.06993	14.3007	0
	Cotang	Tang	Cotang	Tang	Cotang	Tang	Cotang	Tang	′
	89°		88°		87°		86°		

TABLE VII.—*Continued.*

NATURAL TANGENTS AND COTANGENTS.

′	4°		5°		6°		7°		′
	Tang	Cotang	Tang	Cotang	Tang	Cotang	Tang	Cotang	
0	.06993	14.3007	.08749	11.4301	.10510	9.51436	.12278	8.14435	60
1	.07022	14.2411	.08778	11.3919	.10540	9.48781	.12308	8.12481	59
2	.07051	14.1821	.08807	11.3540	.10569	9.46141	.12338	8.10536	58
3	.07080	14.1235	.08837	11.3163	.10599	9.43515	.12367	8.08600	57
4	.07110	14.0655	.08866	11.2789	.10628	9.40904	.12397	8.06674	56
5	.07139	14.0079	.08895	11.2417	.10657	9.38307	.12426	8.04756	55
6	.07168	13.9507	.08925	11.2048	.10687	9.35724	.12456	8.02848	54
7	.07197	13.8940	.08954	11.1681	.10716	9.33155	.12485	8.00948	53
8	.07227	13.8378	.08983	11.1316	.10746	9.30599	.12515	7.99058	52
9	.07256	13.7821	.09013	11.0954	.10775	9.28058	.12544	7.97176	51
10	.07285	13.7267	.09042	11.0594	.10805	9.25530	.12574	7.95302	50
11	.07314	13.6719	.09071	11.0237	.10834	9.23016	.12603	7.93438	49
12	.07344	13.6174	.09101	10.9882	.10863	9.20516	.12633	7.91582	48
13	.07373	13.5634	.09130	10.9529	.10893	9.18028	.12662	7.89734	47
14	.07402	13.5098	.09159	10.9178	.10922	9.15554	.12692	7.87895	46
15	.07431	13.4566	.09189	10.8829	.10952	0.13093	.12722	7.86064	45
16	.07461	13.4039	.09218	10.8483	.10981	9.10646	.12751	7.84242	44
17	.07490	13.3515	.09247	10.8139	.11011	0.08211	.12781	7.82428	43
18	.07519	13.2996	.09277	10.7797	.11040	9.05789	.12810	7.80622	42
19	.07548	13.2480	.09306	10.7457	.11070	9.03379	.12840	7.78825	41
20	.07578	13.1969	.09335	10.7119	.11099	9.00983	.12869	7.77035	40
21	.07607	13.1461	.09365	10.6783	.11128	8.98598	.12899	7.75254	39
22	.07636	13.0958	.09394	10.6450	.11158	8.96227	.12929	7.73480	38
23	.07665	13.0458	.09423	10.6118	.11187	8.93867	.12958	7.71715	37
24	.07695	12.9962	.09453	10.5789	.11217	8.91520	.12988	7.69957	36
25	.07724	12.9469	.09482	10.5462	.11246	8.89185	.13017	7.68208	35
26	.07753	12.8981	.09511	10.5136	.11276	8.86862	.13047	7.66466	34
27	.07782	12.8496	.09541	10.4813	.11305	8.84551	.13076	7.64732	33
28	.07812	12.8014	.09570	10.4491	.11335	8.82252	.13106	7.63005	32
29	.07841	12.7536	.09600	10.4172	.11364	8.79964	.13136	7.61287	31
30	.07870	12.7062	.09629	10.3854	.11394	8.77689	.13165	7.59575	30
31	.07899	12.6591	.09658	10.3538	.11423	8.75425	.13195	7.57872	29
32	.07929	12.6124	.00688	10.3224	.11452	8.73172	.13224	7.56176	28
33	.07958	12.5660	.09717	10.2913	.11482	8.70931	.13254	7.54487	27
34	.07987	12.5199	.09746	10.2602	.11511	8.68701	.13284	7.52806	26
35	.08017	12.4742	.09776	10.2294	.11541	8.66482	.13313	7.51132	25
36	.08046	12.4288	.09805	10.1988	.11570	8.64275	.13343	7.49465	24
37	.08075	12.3838	.09834	10.1683	.11600	8.62078	.13372	7.47806	23
38	.08104	12.3390	.09864	10.1381	.11629	8.59893	.13402	7.46154	22
39	.08134	12.2946	.09893	10.1080	.11659	8.57718	.13432	7.44509	21
40	.08163	12.2505	.09923	10.0780	.11688	8.55555	.13461	7.42871	20
41	.08192	12.2067	.09952	10.0483	.11718	[illegible]	.13491	7.41240	19
42	.08221	12.1632	.09981	10.0187	.11747	8.51259	.13521	7.39616	18
43	.08251	12.1201	.10011	9.98931	.11777	8.49128	.13550	7.37999	17
44	.08280	12.0772	.10040	9.96007	.11806	8.47007	.13580	7.36389	16
45	.08309	12.0346	.10069	9.93101	.11836	8.44896	.13609	7.34786	15
46	.08339	11.9923	.10099	9.90211	.11865	8.42795	.13639	7.33190	14
47	.08368	11.9504	.10128	9.87338	.11895	8.40705	.13669	7.31600	13
48	.08397	11.9087	.10158	9.84482	.11924	8.38625	.13698	7.30018	12
49	.08427	11.8673	.10187	0.81641	.11954	8.36555	.13728	7.28442	11
50	.08456	11.8262	.10216	0.78817	.11983	8.34496	.13758	7.26873	10
51	.08485	11.7853	.10246	8.76009	.12013	8.32446	.13787	7.25310	9
52	.08514	11.7448	.10275	9.73217	.12042	8.30406	.13817	7.23754	8
53	.08544	11.7045	.10305	9.70441	.12072	8.28376	.13846	7.22204	7
54	.08573	11.6645	.10334	9.67680	.12101	8.26355	.13876	7.20661	6
55	.08602	11.6248	.10363	9.64935	.12131	8.24345	.13906	7.19125	5
56	.08632	11.5853	.10393	9.62205	.12160	8.22344	.13935	7.17594	4
57	.08661	11.5461	.10422	9.59490	.12190	8.20352	.13965	7.16071	3
58	.08690	11.5072	.10452	9.56701	.12219	8.18370	.13995	7.14553	2
59	.08720	11.4685	.10481	9.54106	.12249	8.16398	.14024	7.13042	1
60	.08749	11.4301	.10510	9.51436	.12278	8.14435	.14054	7.11537	0
	Cotang	Tang	Cotang	Tang	Cotang	Tang	Cotang	Tang	
′	85°		84°		83°		82°		′

TABLE VII.—*Continued.*

NATURAL TANGENTS AND COTANGENTS.

′	8° Tang	8° Cotang	9° Tang	9° Cotang	10° Tang	10° Cotang	11° Tang	11° Cotang	′
0	.14054	7.11537	.15838	6.31375	.17633	5.67128	.19438	5.14455	60
1	.14084	7.10038	.15868	6.30189	.17663	5.66165	.19468	5.13658	59
2	.14113	7.08546	.15898	6.29007	.17693	5.65205	.19498	5.12862	58
3	.14143	7.07059	.15928	6.27829	.17723	5.64248	.19529	5.12069	57
4	.14173	7.05579	.15958	6.26655	.17753	5.63295	.19559	5.11279	56
5	.14202	7.04105	.15988	6.25486	.17783	5.62344	.19589	5.10490	55
6	.14232	7.02637	.16017	6.24321	.17813	5.61397	.19619	5.09704	54
7	.14262	6.91174	.16047	6.23160	.17843	5.60452	.19649	5.08921	53
8	.14291	6.99718	.16077	6.22003	.17873	5.59511	.19680	5.08139	52
9	.14321	6.98268	.16107	6.20851	.17903	5.58573	.19710	5.07360	51
10	.14351	6.96823	.16137	6.19703	.17933	5.57638	.19740	5.06584	50
11	.14381	6.95385	.16167	6.18559	.17963	5.56706	.19770	5.05809	49
12	.14410	6.93952	.16196	6.17419	.17993	5.55777	.19801	5.05037	48
13	.14440	6.92525	.16226	6.16283	.18023	5.54851	.19831	5.04267	47
14	.14470	6.91104	.16256	6.15151	.18053	5.53927	.19861	5.03499	46
15	.14499	6.89688	.16286	6.14023	.18083	5.53007	.19891	5.02734	45
16	.14529	6.88278	.16316	6.12899	.18113	5.52090	.19921	5.01971	44
17	.14559	6.86874	.16346	6.11779	.18143	5.51176	.19952	5.01210	43
18	.14588	6.85475	.16376	6.10664	.18173	5.50264	.19982	5.00451	42
19	.14618	6.84082	.16405	6.09552	.18203	5.49356	.20012	4.99695	41
20	.14648	6.82694	.16435	6.08444	.18233	5.48451	.20042	4.98940	40
21	.14678	6.81312	.16465	6.07340	.18263	5.47548	.20073	4.98188	39
22	.14707	6.79936	.16495	6.06240	.18293	5.46648	.20103	4.97438	38
23	.14737	6.78564	.16525	6.05143	.18323	5.45751	.20133	4.96690	37
24	.14767	6.77199	.16555	6.04051	.18353	5.44857	.20164	4.95945	36
25	.14796	6.75838	.16585	6.02962	.18384	5.43966	.20194	4.95201	35
26	.14826	6.74483	.16615	6.01878	.18414	5.43077	.20224	4.94460	34
27	.14856	6.73133	.16645	6.00797	.18444	5.42192	.20254	4.93721	33
28	.14886	6.71789	.16674	5.99720	.18474	5.41309	.20285	4.92984	32
29	.14915	6.70450	.16704	5.98646	.18504	5.40429	.20315	4.92249	31
30	.14945	6.69116	.16734	5.97576	.18534	5.39552	.20345	4.91516	30
31	.14975	6.67787	.16764	5.96510	.18564	5.38677	.20376	4.90785	29
32	.15005	6.66463	.16794	5.95448	.18594	5.37805	.20406	4.90056	28
33	.15034	6.65144	.16824	5.94390	.18624	5.36936	.20436	4.89330	27
34	.15064	6.63831	.16854	5.93335	.18654	5.36070	.20466	4.88605	26
35	.15094	6.62523	.16884	5.92283	.18684	5.35206	.20497	4.87882	25
36	.15124	6.61219	.16914	5.91236	.18714	5.34345	.20527	4.87162	24
37	.15153	6.59921	.16944	5.90191	.18745	5.33487	.20557	4.86444	23
38	.15183	6.58627	.16974	5.89151	.18775	5.32631	.20588	4.85727	22
39	.15213	6.57339	.17004	5.88114	.18805	5.31778	.20618	4.85013	21
40	.15243	6.56055	.17033	5.87080	.18835	5.30928	.20648	4.84300	20
41	.15272	6.54777	.17063	5.86051	.18865	5.30080	.20679	4.83590	19
42	.15302	6.53503	.17093	5.85024	.18895	5.29235	.20709	4.82882	18
43	.15332	6.52234	.17123	5.84001	.18925	5.28393	.20739	4.82175	17
44	.15362	6.50970	.17153	5.82982	.18955	5.27553	.20770	4.81471	16
45	.15391	6.49710	.17183	5.81966	.18986	5.26715	.20800	4.80769	15
46	.15421	6.48456	.17213	5.80953	.19016	5.25880	.20830	4.80068	14
47	.15451	6.47206	.17243	5.79944	.19046	5.25048	.20861	4.79370	13
48	.15481	6.45961	.17273	5.78938	.19076	5.24218	.20891	4.78673	12
49	.15511	6.44720	.17303	5.77936	.19106	5.23391	.20921	4.77978	11
50	.15540	6.43484	.17333	5.76937	.19136	5.22566	.20952	4.77286	10
51	.15570	6.42253	.17363	5.75941	.19166	5.21744	.20982	4.76595	9
52	.15600	6.41026	.17393	5.74949	.19197	5.20925	.21013	4.75906	8
53	.15630	6.39804	.17423	5.73960	.19227	5.20107	.21043	4.75219	7
54	.15660	6.38587	.17453	5.72974	.19257	5.19293	.21073	4.74534	6
55	.15689	6.37374	.17483	5.71992	.19287	5.18480	.21104	4.73851	5
56	.15719	6.36165	.17513	5.71013	.19317	5.17671	.21134	4.73170	4
57	.15749	6.34961	.17543	5.70037	.19347	5.16863	.21164	4.72490	3
58	.15779	6.33761	.17573	5.69064	.19378	5.16058	.21195	4.71813	2
59	.15809	6.32566	.17603	5.68094	.19408	5.15256	.21225	4.71137	1
60	.15838	6.31375	.17633	5.67128	.19438	5.14455	.21256	4.70463	0
′	81° Cotang	81° Tang	80° Cotang	80° Tang	79° Cotang	79° Tang	78° Cotang	78° Tang	′

TABLE VII.—*Continued.*

NATURAL TANGENTS AND COTANGENTS.

′	12°		13°		14°		15°		′
	Tang	Cotang	Tang	Cotang	Tang	Cotang	Tang	Cotang	
0	.21256	4.70463	.23087	4.33148	.24933	4.01078	.26795	3.73205	60
1	.21286	4.69791	.23117	4.32573	.24964	4.00582	.26826	3.72771	59
2	.21316	4.69121	.23148	4.32001	.24995	4.00086	.26857	3.72338	58
3	.21347	4.68452	.23179	4.31430	.25026	3.99592	.26888	3.71907	57
4	.21377	4.67786	.23209	4.30860	.25056	3.99099	.26920	3.71476	56
5	.21408	4.67121	.23240	4.30291	.25087	3.98607	.26951	3.71046	55
6	.21438	4.66458	.23271	4.29724	.25118	3.98117	.26982	3.70616	54
7	.21469	4.65797	.23301	4.29159	.25149	3.97627	.27013	3.70188	53
8	.21499	4.65138	.23332	4.28595	.25180	3.97139	.27044	3.69761	52
9	.21529	4.64480	.23363	4.28032	.25211	3.96651	.27076	3.69335	51
10	.21560	4.63825	.23393	4.27471	.25242	3.96165	.27107	3.68909	50
11	.21590	4.63171	.23424	4.26911	.25273	3.95680	.27138	3.68485	49
12	.21621	4.62518	.23455	4.26352	.25304	3.95196	.27169	3.68061	48
13	.21651	4.61868	.23485	4.25795	.25335	3.94713	.27201	3.67638	47
14	.21682	4.61219	.23516	4.25239	.25366	3.94232	.27232	3.67217	46
15	.21712	4.60572	.23547	4.24685	.25397	3.93751	.27263	3.66796	45
16	.21743	4.59927	.23578	4.24132	.25428	3.93271	.27294	3.66376	44
17	.21773	4.59283	.23608	4.23580	.25459	3.92793	.27326	3.65957	43
18	.21804	4.58641	.23639	4.23030	.25490	3.92316	.27357	3.65538	42
19	.21834	4.58001	.23670	4.22481	.25521	3.91839	.27388	3.65121	41
20	.21864	4.57363	.23700	4.21933	.25552	3.91364	.27419	3.64705	40
21	.21895	4.56726	.23731	4.21387	.25583	3.90890	.27451	3.64289	39
22	.21925	4.56091	.23762	4.20842	.25614	3.90417	.27482	3.63874	38
23	.21956	4.55458	.23793	4.20298	.25645	3.89945	.27513	3.63461	37
24	.21986	4.54826	.23823	4.19756	.25676	3.89474	.27545	3.63048	36
25	.22017	4.54196	.23854	4.19215	.25707	3.89004	.27576	3.62636	35
26	.22047	4.53568	.23885	4.18675	.25738	3.88536	.27607	3.62224	34
27	.22078	4.52941	.23916	4.18137	.25769	3.88068	.27638	3.61814	33
28	.22108	4.52316	.23946	4.17600	.25800	3.87601	.27670	3.61405	32
29	.22139	4.51693	.23977	4.17064	.25831	3.87136	.27701	3.60996	31
30	.22169	4.51071	.24008	4.16530	.25862	3.86671	.27732	3 60588	30
31	.22200	4.50451	.24039	4.15997	.25893	3.86208	.27764	3.60181	29
32	.22231	4.49832	.24069	4.15465	.25924	3.85745	.27795	3.59775	28
33	.22261	4.49215	.24100	4.14934	.25955	3.85284	.27826	3.59370	27
34	.22292	4.48600	.24131	4.14405	.25986	3.84824	.27858	3.58966	26
35	.22322	4.47986	.24162	4.13877	.20017	3.84364	.27889	3.58562	25
36	.22353	4.47374	.24193	4.13350	.26048	3.83906	.27921	3.58160	24
37	.22383	4.46764	.24223	4.12825	.26079	3.83449	.27952	3.57758	23
38	.22414	4.46155	.24254	4.12301	.26110	3.82992	.27983	3.57357	22
39	.22444	4.45548	.24285	4.11778	.26141	3.82537	.28015	3.56957	21
40	.22475	4.44942	.24316	4.11256	.26172	3.82083	.28046	3.56557	20
41	.22505	4.44338	.24347	4.10736	.26203	3.81630	.28077	3.56159	19
42	.22536	4.43735	.24377	4.10216	.26235	3.81177	.28109	3.55761	18
43	.22567	4.43134	.24408	4.09699	.26266	3.80726	.28140	3.55364	17
44	.22597	4.42534	.24439	4.09182	.26297	3.80276	.28172	3.54968	16
45	.22628	4.41936	.24470	4.08666	.26328	3.79827	.28203	3.54573	15
46	.22658	4.41340	.24501	4.08152	.26359	3.79378	.28234	3.54179	14
47	.22689	4.40745	.24532	4.07639	.26390	3.78931	.28266	3.53785	13
48	.22719	4.40152	.24562	4.07127	.26421	3.78485	.28297	3.53393	12
49	.22750	4.39560	.24593	4.06616	.26452	3.78040	.28329	3.53001	11
50	.22781	4.38969	.24624	4.06107	.26483	3.77595	.28360	3.52609	10
51	.22811	4.38381	.24655	4.05599	.26515	3.77152	.28391	3.52219	9
52	.22842	4.37793	.24686	4.05092	.26546	3.76709	.28423	3.51829	8
53	.22872	4.37207	.24717	4.04586	.26577	3.76268	.28454	3.51441	7
54	.22903	4.36623	.24747	4.04081	.26608	3.75828	.28486	3.51053	6
55	.22934	4.36040	.24778	4.03578	.26639	3.75388	.28517	3.50666	5
56	.22964	4.35459	.24809	4.03076	.26670	3.74950	.28549	3.50279	4
57	.22995	4.34879	.24840	4.02574	.26701	3.74512	.28580	3.49894	3
58	.23026	4.34300	.24871	4.02074	.26733	3.74075	.28612	3.49509	2
59	.23056	4.33723	.24902	4.01576	.26764	3.73640	.28643	3.49125	1
60	.23087	4.33148	.24933	4.01078	.26795	3.73205	.28675	3.48741	0
′	Cotang	Tang	Cotang	Tang	Cotang	Tang	Cotang	Tang	′
	77°		76°		75°		74		

TABLE VII.—*Continued.*

NATURAL TANGENTS AND COTANGENTS.

′	16° Tang	16° Cotang	17° Tang	17° Cotang	18° Tang	18° Cotang	19° Tang	19° Cotang	′
0	.28675	3.48741	.30573	3.27085	.32492	3.07768	.34433	2.90421	60
1	.28706	3.48359	.30605	3.26745	.32524	3.07464	.34465	2.90147	59
2	.28738	3.47977	.30637	3.26406	.32556	3.07160	.34498	2.89873	58
3	.28769	3.47596	.30669	3.26067	.32588	3.06857	.34530	2.89600	57
4	.28800	3.47216	.30700	3.25729	.32621	3.06554	.34563	2.89327	56
5	.28832	3.46837	.30732	3.25392	.32653	3.06252	.34596	2.89055	55
6	.28864	3.46458	.30764	3.25055	.32685	3.05950	.34628	2.88783	54
7	.28895	3.46080	.30796	3.24719	.32717	3.05649	.34661	2.88511	53
8	.28927	3.45703	.30828	3.24383	.32749	3.05349	.34693	2.88240	52
9	.28958	3.45327	.30860	3.24049	.32782	3.05049	.34726	2.87970	51
10	.28990	3.44951	.30891	3.23714	.32814	3.04749	.34758	2 87700	50
11	.29021	3.44576	.30923	3.23381	.32846	3.04450	.34791	2.87430	49
12	.29053	3.44202	.30955	3.23048	.32878	3.04152	.34824	2.87161	48
13	.29084	3.43829	.30987	3.22715	.32911	3.03854	.34856	2.86892	47
14	.29116	3.43456	.31019	3.22384	.32943	3.03556	.34889	2.86624	46
15	.29147	3.43084	.31051	3.22053	.32975	3.03260	.34922	2.86356	45
16	.29179	3.42713	.31083	3.21722	.33007	3.02963	.34954	2.86089	44
17	.29210	3.42343	.31115	3.21392	.33040	3.02667	.34987	2.85822	43
18	.29242	3.41973	.31147	3.21063	.33072	3.02372	.35020	2.85555	42
19	.29274	3.41604	.31178	3.20734	.33104	3.02077	.35052	2.85289	41
20	.29305	3.41236	.31210	3.20406	.33136	3.01783	.35085	2.85023	40
21	.29337	3.40869	.31242	3.20079	.33169	3.01489	.35118	2.84758	39
22	.29368	3.40502	.31274	3.19752	.33201	3.01196	.35150	2.84494	38
23	.29400	3.40136	.31306	3.19426	.33233	3.00903	.35183	2.84229	37
24	.29432	3.39771	.31338	3.19100	.33266	3.00611	.35216	2.83965	36
25	.29463	3.39406	.31370	3.18775	.33298	3.00319	.35248	2.83702	35
26	.29495	3.39042	.31402	3.18451	.33330	3.00028	.35281	2.83439	34
27	.29526	3.38679	.31434	3.18127	.33363	2.99738	.35314	2.83176	33
28	.29558	3.38317	.31466	3.17804	.33395	2.99447	.35346	2.82914	32
29	.29590	3.37955	.31498	3.17481	.33427	2.99158	.35379	2.82653	31
30	.29621	3.37594	.31530	3.17159	.33460	2.98868	.35412	2.82391	30
31	.29653	3.37234	.31562	3.16838	.33492	2.98580	.35445	2.82130	29
32	.29685	3.36875	.31594	3.16517	.33524	2.98292	.35477	2.81870	28
33	.29716	3.36516	.31626	3.16197	.33557	2.98004	.35510	2.81610	27
34	.29748	3.36158	.31658	3.15877	.33589	2.97717	.35543	2.81350	26
35	.29780	3.35800	.31690	3.15558	.33621	2.97430	.35576	2.81091	25
36	.29811	3.35443	.31722	3.15240	.33654	2.97144	.35608	2.80833	24
37	.29843	3.35087	.31754	3.14923	.33686	2.96858	.35641	2.80574	23
38	.29875	3.34732	.31786	3.14605	.33718	2.96573	.35674	2.80316	22
39	.29906	3.34377	.31818	3.14288	.33751	2.96288	.35707	2.80059	21
40	.29938	3.34023	.31850	3.13972	.33783	2.96004	.35740	2.79802	20
41	.29970	3.33670	.31882	3.13656	.33816	2.95721	.35772	2.79545	19
42	.30001	3.33317	.31914	3.13341	.33848	2.95437	.35805	2.79289	18
43	.30033	3.32965	.31946	3.13027	.33881	2.95155	.35838	2.79033	17
44	.30065	3.32614	.31978	3.12713	.33913	2.94872	.35871	2.78778	16
45	.30097	3.32264	.32010	3.12400	.33945	2.94591	.35904	2.78523	15
46	.30128	3.31914	.32042	3.12087	.33978	2.94309	.35937	2.78269	14
47	.30160	3.31565	.32074	3.11775	.34010	2.94028	.35969	2.78014	13
48	.30192	3.31216	.22106	3.11464	.34043	2.93748	.36002	2.77761	12
49	.30224	3.30868	.32139	3.11153	.34075	2.93468	.36035	2.77507	11
50	.30255	3.30521	.32171	3.10842	.34108	2.93189	.36068	2.77254	10
51	.30287	3.30174	.32203	3.10532	.34140	2.92910	.36101	2.77002	9
52	.30319	3.29829	.32235	3.10223	.34173	2.92632	.36134	2.76750	8
53	.30351	3.29483	.32267	3.09914	.34205	2.92354	.36167	2.76498	7
54	.30382	3.29139	.32299	3.09606	.34238	2.92076	.36199	2.76247	6
55	.30414	3.28795	.32331	3.09298	.34270	2.91799	.36232	2.75996	5
56	.30446	3.28452	.32363	3.08991	.34303	2.91523	.36265	2.75746	4
57	.30478	3.28109	.32396	3.08685	.34335	2.91246	.36298	2.75496	3
58	.30509	3.27767	.32428	3.08379	.34368	2.90971	.36331	2.75246	2
59	.30541	3.27426	.32460	3.08073	.34400	2.90696	.36364	2.74997	1
60	.30573	3.27085	.32492	3.07768	.34433	2.90421	.36397	2.74748	0
′	73° Cotang	73° Tang	72° Cotang	72° Tang	71° Cotang	71° Tang	70° Cotang	70° Tang	′

TABLE VII.—*Continued.*

NATURAL TANGENTS AND COTANGENTS.

′	20°		21°		22°		23°		′
	Tang	Cotang	Tang	Cotang	Tang	Cotang	Tang	Cotang	
0	.36397	2.74748	.38386	2.60509	.40403	2.47509	.42447	2.35585	60
1	.36430	2.74499	.38420	2.60283	.40436	2.47302	.42482	2.35395	59
2	.36463	2.74251	.38453	2.60057	.40470	2.47095	.42516	2.35205	58
3	.36496	2.74004	.38487	2.59831	.40504	2.46888	.42551	2.35015	57
4	.36529	2.73756	.38520	2.59606	.40538	2.46682	.42585	2.34825	56
5	.36562	2.73509	.38553	2.59381	.40572	2.46476	.42619	2.34636	55
6	.36595	2.73263	.38587	2.59156	.40606	2.46270	.42654	2.34447	54
7	.36628	2.73017	.38620	2.58932	.40640	2.46065	.42688	2.34258	53
8	.36661	2.72771	.38654	2.58708	.40674	2.45860	.42722	2.34069	52
9	.36694	2.72526	.38687	2.58484	.40707	2.45655	.42757	2.33881	51
10	.36727	2.72281	.38721	2.58261	.40741	2.45451	.42791	2.33693	50
11	.36760	2.72036	.38754	2.58038	.40775	2.45246	.42826	2.33505	49
12	.36793	2.71792	.38787	2.57815	.40809	2.45043	.42860	2.33317	48
13	.36826	2.71548	.38821	2.57593	.40843	2.44839	.42894	2.33130	47
14	.36859	2.71305	.38854	2.57371	.40877	2.44636	.42929	2.32943	46
15	.36892	2.71062	.38888	2.57150	.40911	2.44433	.42963	2.32756	45
16	.36925	2.70819	.38921	2.56928	.40945	2.44230	.42998	2.32570	44
17	.36958	2.70577	.38955	2.56707	.40979	2.44027	.43032	2.32383	43
18	.36991	2.70335	.38988	2.56487	.41013	2.43825	.43067	2.32197	42
19	.37024	2.70094	.39022	2.56266	.41047	2.43623	.43101	2.32012	41
20	.37057	2.69853	.39055	2.56046	.41081	2.43422	.43136	2.31826	40
21	.37090	2.69612	.39089	2.55827	.41115	2.43220	.43170	2.31641	39
22	.37123	2.69371	.39122	2.55608	.41149	2.43019	.43205	2.31456	38
23	.37157	2.69131	.39156	2.55389	.41183	2.42819	.43239	2.31271	37
24	.37190	2.68892	.39190	2.55170	.41217	2.42618	.43274	2.31086	36
25	.37223	2.68653	.39223	2.54952	.41251	2.42418	.43308	2.30902	35
26	.37256	2.68414	.39257	2.54734	.41285	2.42218	.43343	2.30718	34
27	.37289	2.68175	.39290	2.54516	.41319	2.42019	.43378	2.30534	33
28	.37322	2.67937	.39324	2.54299	.41353	2.41819	.43412	2.30351	32
29	.37355	2.67700	.39357	2.54082	.41387	2.41620	.43447	2.30167	31
30	.37388	2.67462	.39391	2.53865	.41421	2.41421	.43481	2.29984	30
31	.37422	2.67225	.39425	2.53648	.41455	2.41223	.43516	2.29801	29
32	.37455	2.66989	.39458	2.53432	.41490	2.41025	.43550	2.29619	28
33	.37488	2.66752	.39492	2.53217	.41524	2.40827	.43585	2.29437	27
34	.37521	2.66516	.39526	2.53001	.41558	2.40629	.43620	2.29254	26
35	.37554	2.66281	.39559	2.52786	.41592	2.40432	.43654	2.29073	25
36	.37588	2.66046	.39593	2.52571	.41626	2.40235	.43689	2.28891	24
37	.37621	2.65811	.39626	2.52357	.41660	2.40038	.43724	2.28710	23
38	.37654	2.65576	.39660	2.52142	.41694	2.39841	.43758	2.28528	22
39	.37687	2.65342	.39694	2.51929	.41728	2.39645	.43793	2.28348	21
40	.37720	2.65109	.39727	2.51715	.41763	2.39449	.43828	2.28167	20
41	.37754	2.64875	.39761	2.51502	.41797	2.39253	.43862	2.27987	19
42	.37787	2.64642	.39795	2.51289	.41831	2.39058	.43897	2.27806	18
43	.37820	2.64410	.39829	2.51076	.41865	2.38863	.43932	2.27626	17
44	.37853	2.64177	.39862	2.50864	41899	2.38668	.43966	2.27447	16
45	.37887	2.63945	.39896	2.50652	.41933	2.38473	.44001	2.27267	15
46	.37920	2.63714	.39930	2.50440	.41968	2.38279	.44036	2.27088	14
47	.37953	2.63483	.39963	2.50229	.42002	2.38084	.44071	2.26909	13
48	.37986	2.63252	.39997	2.50018	.42036	2.37891	.44105	2.26730	12
49	.38020	2.63021	.40031	2.49807	.42070	2.37697	.44140	2.26552	11
50	.38053	2.62791	.40065	2.49597	.42105	2.37504	.44175	2.26374	10
51	.38086	2.62561	.40098	2.49386	.42139	2.37311	.44210	2.26196	9
52	.38120	2.62332	.40132	2.49177	.42173	2.37118	.44244	2.26018	8
53	.38153	2.62103	.40166	2.48967	.42207	2.36925	.44279	2.25840	7
54	.38186	2.61874	.40200	2.48758	.42242	2.36733	.44314	2.25663	6
55	.38220	2.61646	.40234	2.48549	.42276	2.36541	.44349	2.25486	5
56	.38253	2.61418	.40267	2 48340	.42310	2.36349	.44384	2.25309	4
57	.38286	2.61190	.40301	2.48132	.42345	2.36158	.44418	2.25132	3
58	.38320	2.60963	.40335	2.47924	.42379	2.35967	.44453	2.24956	2
59	.38353	2.60736	.40369	2.47716	.42413	2.35776	.44488	2.24780	1
60	.38386	2.60509	.40403	2.47509	.42447	2.35585	.44523	2.24604	0
′	Cotang	Tang	Cotang	Tang	Cotang	Tang	Cotang	Tang	′
	69°		68°		67°		66°		

TABLE VII.—*Continued.*

NATURAL TANGENTS AND COTANGENTS.

′	24°		25°		26°		27°		′
	Tang	Cotang	Tang	Cotang	Tang	Cotang	Tang	Cotang	
0	.44523	2.24604	.46631	2.14451	.48773	2.05030	.50953	1.96261	60
1	.44558	2.24428	.46666	2.14288	.48809	2.04879	.50989	1.96120	59
2	.44593	2.24252	.46702	2.14125	.48845	2.04728	.51026	1.95979	58
3	.44627	2.24077	.46737	2.13963	.48881	2.04577	.51063	1.95838	57
4	.44662	2.23902	.46772	2.13801	.48917	2.04426	.51099	1.95698	56
5	.44697	2.23727	.46808	2.13639	.48953	2.04276	.51136	1.95557	55
6	.44732	2.23553	.46843	2.13477	.48989	2.04125	.51173	1.95417	54
7	.44767	2.23378	.46879	2.13316	.49026	2.03975	.51209	1.95277	53
8	.44802	2.23204	.46914	2.13154	.49062	2.03825	.51246	1.95137	52
9	.44837	2.23030	.46950	2.12993	.49098	2.03675	.51283	1.94997	51
10	.44872	2.22857	.46985	2.12832	.49134	2.03526	.51319	1.94858	50
11	.44907	2.22683	.47021	2.12671	.49170	2.03376	.51356	1.94718	49
12	.44942	2.22510	.47056	2.12511	.49206	2.03227	.51393	1.94579	48
13	.44977	2.22337	.47092	2.12350	.49242	2.03078	.51430	1.94440	47
14	.45012	2.22164	.47128	2.12190	.49278	2.02929	.51467	1.94301	46
15	.45047	2.21992	.47163	2.12030	.49315	2.02780	.51503	1.94162	45
16	.45082	2.21819	.47199	2.11871	.49351	2.02631	.51540	1.94023	44
17	.45117	2.21647	.47234	2.11711	.49387	2.02483	.51577	1.93885	43
18	.45152	2.21475	.47270	2.11552	.49423	2.02335	.51614	1.93746	42
19	.45187	2.21304	.47305	2.11392	.49459	2.02187	.51651	1.93608	41
20	.45222	2.21132	.47341	2.11233	.49495	2.02039	.51688	1.93470	40
21	.45257	2.20961	.47377	2.11075	.49532	2.01891	.51724	1.93332	39
22	.45292	2.20790	.47412	2.10916	.49568	2.01743	.51761	1.93195	38
23	.45327	2.20619	.47448	2.10758	.49604	2.01596	.51798	1.93057	37
24	.45362	2.20449	.47483	2.10600	.49640	2.01449	.51835	1.92920	36
25	.45397	2.20278	.47519	2.10442	.49677	2.01302	.51872	1.92782	35
26	.45432	2.20108	.47555	2.10284	.49713	2.01155	.51909	1.92645	34
27	.45467	2.19938	.47590	2.10126	.49749	2.01008	.51946	1.92508	33
28	.45502	2.19769	.47626	2.09969	.49786	2.00862	.51983	1.92371	32
29	.45538	2.19599	.47662	2.09811	.49822	2.00715	.52020	1.92235	31
30	.45573	2.19430	.47698	2.09654	.49858	2.00569	.52057	1.92098	30
31	.45608	2.19261	.47733	2.09498	.49894	2.00423	.52094	1.91962	29
32	.45643	2.19092	.47769	2.09341	.49931	2.00277	.52131	1.91826	28
33	.45678	2.18923	.47805	2.09184	.49967	2.00131	.52168	1.91690	27
34	.45713	2.18755	.47840	2.09028	.50004	1.99986	.52205	1.91554	26
35	.45748	2.18587	.47876	2.08872	.50040	1.99841	.52242	1.91418	25
36	.45784	2.18419	.47912	2.08716	.50076	1.99695	.52279	1.91282	24
37	.45819	2.18251	.47948	2.08560	.50113	1.99550	.52316	1.91147	23
38	.45854	2.18084	.47984	2.08405	.50149	1.99406	.52353	1.91012	22
39	.45889	2.17916	.48019	2.08250	.50185	1.99261	.52390	1.90876	21
40	.45924	2.17749	.48055	2.08094	.50222	1.99116	.52427	1.90741	20
41	.45960	2.17582	.48091	2.07939	.50258	1.98972	.52464	1.90607	19
42	.45995	2.17416	.48127	2.07785	.50295	1.98828	.52501	1.90472	18
43	.46030	2.17249	.48163	2.07630	.50331	1.98684	.52538	1.90337	17
44	.46065	2.17083	.48198	2.07476	.50368	1.98540	.52575	1.90203	16
45	.46101	2.16917	.48234	2.07321	.50404	1.98396	.52613	1.90069	15
46	.46136	2.16751	.48270	2.07167	.50441	1.98253	.52650	1.89935	14
47	.46171	2.16585	.48306	2.07014	.50477	1.98110	.52687	1.89801	13
48	.46206	2.16420	.48342	2.06860	.50514	1.97966	.52724	1.89667	12
49	.46242	2.16255	.48378	2.06706	.50550	1.97823	.52761	1.89533	11
50	.46277	2.16090	.48414	2.06553	.50587	1.97681	.52798	1.89400	10
51	.46312	2.15925	.48450	2.06400	.50623	1.97538	.52836	1.89266	9
52	.46348	2.15760	.48486	2.06247	.50660	1.97395	.52873	1.89133	8
53	.46383	2.15596	.48521	2.06094	.50696	1.97253	.52910	1.89000	7
54	.46418	2.15432	.48557	2.05942	.50733	1.97111	.52947	1.88867	6
55	.46454	2.15268	.48593	2.05790	.50769	1.96969	.52985	1.88734	5
56	.46489	2.15104	.48629	2.05637	.50806	1.96827	.53022	1.88602	4
57	.46525	2.14940	.48665	2.05485	.50843	1.96685	.53059	1.88469	3
58	.46560	2.14777	.48701	2.05333	.50879	1.96544	.53096	1.88337	2
59	.46595	2.14614	.48737	2.05182	.50916	1.96402	.53134	1.88205	1
60	.46631	2.14451	.48773	2.05030	.50953	1.96261	.53171	1.88073	0
′	Cotang	Tang	Cotang	Tang	Cotang	Tang	Cotang	Tang	′
	65°		64°		63°		62°		

TABLE VII.—*Continued.*

NATURAL TANGENTS AND COTANGENTS.

′	28°		29°		30°		31°		′
	Tang	Cotang	Tang	Cotang	Tang	Cotang	Tang	Cotang	
0	.53171	1.88073	.55431	1.80405	.57735	1.73205	.60086	1.66428	60
1	.53208	1.87941	.55469	1.80281	.57774	1.73089	.60126	1.66318	59
2	.53246	1.87809	.55507	1.80158	.57813	1.72973	.60165	1.66209	58
3	.53283	1.87677	.55545	1.80034	.57851	1.72857	.60205	1.66099	57
4	.53320	1.87546	.55583	1.79911	.57890	1.72741	.60245	1.65990	56
5	.53358	1.87415	.55621	1.79788	.57929	1.72625	.60284	1.65881	55
6	.53395	1.87283	.55659	1.79665	.57968	1.72509	.60324	1.65772	54
7	.53432	1.87152	.55697	1.79542	.58007	1.72393	.60364	1.65663	53
8	.53470	1.87021	.55736	1.79419	.58046	1.72278	.60403	1.65554	52
9	.53507	1.86891	.55774	1.79296	.58085	1.72163	.60443	1.65445	51
10	.53545	1.86760	.55812	1.79174	.58124	1.72047	.60483	1.65337	50
11	.53582	1.86630	.55850	1.79051	.58162	1.71932	.60522	1.65228	49
12	.53620	1.86499	.55888	1.78929	.58201	1.71817	.60562	1.65120	48
13	.53657	1.86369	.55926	1.78807	.58240	1.71702	.60602	1.65011	47
14	.53694	1.86239	.55964	1.78685	.58279	1.71588	.60642	1.64903	46
15	.53732	1.86109	.56003	1.78563	.58318	1.71473	.60681	1.64795	45
16	.53769	1.85979	.56041	1.78441	.58357	1.71358	.60721	1.64687	44
17	.53807	1.85850	.56079	1.78319	.58396	1.71244	.60761	1.64579	43
18	.53844	1.85720	.56117	1.78198	.58435	1.71129	.60801	1.64471	42
19	.53882	1.85591	.56156	1.78077	.58474	1.71015	.60841	1.64363	41
20	.53920	1.85462	.56194	1.77955	.58513	1.70901	.60881	1.64256	40
21	.53957	1.85333	.56232	1.77834	.58552	1.70787	.60921	1.64148	39
22	.53995	1.85204	.56270	1.77713	.58591	1.70673	.60960	1.64041	38
23	.54032	1.85075	.56309	1.77592	.58631	1.70560	.61000	1.63934	37
24	.54070	1.84946	.56347	1.77471	.58670	1.70446	.61040	1.63826	36
25	.54107	1.84818	.56385	1.77351	.58709	1.70332	.61080	1.63719	35
26	.54145	1.84689	.56424	1.77230	.58748	1.70219	.61120	1.63612	34
27	.54183	1.84561	.56462	1.77110	.58787	1.70106	.61160	1.63505	33
28	.54220	1.84433	.56501	1.76990	.58826	1.69992	.61200	1.63398	32
29	.51258	1.84305	.56539	1.76869	.58865	1.69879	.61240	1.63292	31
30	.54296	1.84177	.56577	1.76749	.58905	1.69766	.61280	1.63185	30
31	.54333	1.84049	.56616	1.76629	.58044	1.69653	.61320	1.63079	29
32	.54371	1.83922	.56654	1.76510	.58983	1.69541	.61360	1.62972	28
33	.54409	1.83794	.56693	1.76390	.59022	1.69428	.61400	1.62866	27
34	.54446	1.83667	.56731	1.76271	.59061	1.69316	.61440	1.62760	26
35	.54484	1.83540	.56769	1.76151	.59101	1.69203	.61480	1.62654	25
36	.54522	1.83413	.56808	1.76032	.59140	1.69091	.61520	1.62548	24
37	.54560	1.83286	.56846	1.75913	.59179	1.68979	.61561	1.62442	23
38	.54597	1.83159	.56885	1 75794	.59218	1.68866	.61601	1.62336	22
39	.54635	1.83033	.56923	1.75675	.59258	1.68754	.61641	1.62230	21
40	.54673	1.82906	.56962	1.75556	.59297	1.68643	.61681	1.62125	20
41	.54711	1.82780	.57000	1.75437	.59336	1.68531	.61721	1.62019	19
42	.54748	1.82654	.57039	1.75319	.59376	1.68419	.61761	1.61914	18
43	.54786	1.82528	.57078	1.75200	.59415	1.68308	.61801	1.61808	17
44	.54824	1.82402	.57116	1.75082	.59454	1.68196	.61842	1.61703	16
45	.54862	1.82276	.57155	1.74964	.59494	1.68085	.61882	1.61598	15
46	.54900	1.82150	.57193	1.74846	.59533	1.67974	.61922	1.61493	14
47	.54938	1.82025	.57232	1.74728	.59573	1.67863	.61962	1.61388	13
48	.54975	1.81899	.57271	1.74610	.59612	1.67752	.62003	1.61283	12
49	.55013	1.81774	.57309	1.74492	.59651	1.67641	.62043	1.61179	11
50	.55051	1.81649	.57348	1.74375	.59691	1.67530	.62083	1.61074	10
51	.55089	1.81524	.57386	1.74257	.59730	1.67419	.62124	1.60970	9
52	.55127	1.81399	.57425	1.74140	.59770	1.67309	.62164	1.60865	8
53	.55165	1.81274	.57464	1.74022	.59809	1.67198	.62204	1.60761	7
54	.55203	1.81150	.57503	1.73905	.59849	1.67088	.62245	1.60657	6
55	.55241	1.81025	.57541	1.73788	.59888	1.66978	.62285	1.60553	5
56	.55279	1.80901	.57580	1.73671	.59928	1.66867	.62325	1.60449	4
57	.55317	1.80777	.57619	1.73555	.59967	1.66757	.62366	1.60345	3
58	.55355	1.80653	.57657	1.73438	.60007	1.66647	.62406	1.60241	2
59	.55393	1.80529	.57696	1.73321	.60046	1.66538	.62446	1.60137	1
60	.55431	1.80405	.57735	1.73205	.60086	1.66428	.62487	1.60033	0
′	Cotang	Tang	Cotang	Tang	Cotang	Tang	Cotang	Tang	′
	61°		60°		59°		58°		

TABLE VII.—*Continued.*

NATURAL TANGENTS AND COTANGENTS.

′	32° Tang	32° Cotang	33° Tang	33° Cotang	34° Tang	34° Cotang	35° Tang	35° Cotang	′
0	.62487	1.60033	.64941	1.53986	.67451	1.48256	.70021	1.42815	60
1	.62527	1.59930	.64982	1.53888	.67493	1.48163	.70064	1.42726	59
2	.62568	1.59826	.65024	1.53791	.67536	1.48070	.70107	1.42638	58
3	.62608	1.59723	.65065	1.53693	.67578	1.47977	.70151	1.42550	57
4	.62649	1.59620	.65106	1.53595	.67620	1.47885	.70194	1.42462	56
5	.62689	1.59517	.65148	1.53497	.67663	1.47792	.70238	1.42374	55
6	.62730	1.59414	.65189	1.53400	.67705	1.47699	.70281	1.42286	54
7	.62770	1.59311	.65231	1.53302	.67748	1.47607	.70325	1.42198	53
8	.62811	1.59208	.65272	1.53205	.67790	1.47514	.70368	1.42110	52
9	.62852	1.59105	.65314	1.53107	.67832	1.47422	.70412	1.42022	51
10	.62892	1.59002	.65355	1.53010	.67875	1.47330	.70455	1.41934	50
11	.62933	1.58900	.65397	1.52913	.67917	1.47238	.70499	1.41847	49
12	.62973	1.58797	.65438	1.52816	.67960	1.47146	.70542	1.41759	48
13	.63014	1.58695	.65480	1.52719	.68002	1.47053	.70586	1.41672	47
14	.63055	1.58593	.65521	1.52622	.68045	1.46962	.70629	1.41584	46
15	.63095	1.58490	.65563	1.52525	.68088	1.46870	.70673	1.41497	45
16	.63136	1.58388	.65604	1.52429	.68130	1.46778	.70717	1.41409	44
17	.63177	1.58286	.65646	1.52332	.68173	1.46686	.70760	1.41322	43
18	.63217	1.58184	.65688	1.52235	.68215	1.46595	.70804	1.41235	42
19	.63258	1.58083	.65729	1.52139	.68258	1.46503	.70848	1.41148	41
20	.63299	1.57981	.65771	1.52043	.68301	1.46411	.70891	1.41061	40
21	.63340	1.57879	.65813	1.51946	.68343	1.46320	.70935	1.40974	39
22	.63380	1.57778	.65854	1.51850	.68386	1.46229	.70979	1.40887	38
23	.63421	1.57676	.65896	1.51754	.68429	1.46137	.71023	1.40800	37
24	.63462	1.57575	.65938	1.51658	.68471	1.46046	.71066	1.40714	36
25	.63503	1.57474	.65980	1.51562	.68514	1.45955	.71110	1.40627	35
26	.63544	1.57372	.66021	1.51466	.68557	1.45864	.71154	1.40540	34
27	.63584	1.57271	.66063	1.51370	.68600	1.45773	.71198	1.40454	33
28	.63625	1.57170	.66105	1.51275	.68642	1.45682	.71242	1.40367	32
29	.63666	1.57069	.66147	1.51179	.68685	1.45592	.71285	1.40281	31
30	.63707	1.56969	.66189	1.51084	.68728	1.45501	.71329	1.40195	30
31	.63748	1.56868	.66230	1.50988	.68771	1.45410	.71373	1.40109	29
32	.63789	1.56767	.66272	1.50893	.68814	1.45320	.71417	1.40022	28
33	.63830	1.56667	.66314	1.50797	.68857	1.45229	.71461	1.39936	27
34	.63871	1.56566	.66356	1.50702	.68900	1.45139	.71505	1.39850	26
35	.63912	1.56466	.66398	1.50607	.68942	1.45049	.71549	1.39764	25
36	.63953	1.56366	.66440	1.50512	.68985	1.44958	.71593	1.39679	24
37	.63994	1.56265	.66482	1.50417	.69028	1.44868	.71637	1.39593	23
38	.64035	1.56165	.66524	1.50322	.69071	1.44778	.71681	1.39507	22
39	.64076	1.56065	.66566	1.50228	.69114	1.44688	.71725	1.39421	21
40	.64117	1.55966	.66608	1.50133	.69157	1.44598	.71769	1.39336	20
41	.64158	1.55866	.66650	1.50038	.69200	1.44508	.71813	1.39250	19
42	.64199	1.55766	.66692	1.49944	.69243	1.44418	.71857	1.39165	18
43	.64240	1.55666	.66734	1.49849	.69286	1.44329	.71901	1.39079	17
44	.64281	1.55567	.66776	1.49755	.69329	1.44239	.71946	1.38994	16
45	.64322	1.55467	.66818	1.49661	.69372	1.44149	.71990	1.38909	15
46	.64363	1.55368	.66860	1.49566	.69416	1.44060	.72034	1.38824	14
47	.64404	1.55269	.66902	1.49472	.69459	1.43970	.72078	1.38738	13
48	.64446	1.55170	.66944	1.49378	.69502	1.43881	.72122	1.38653	12
49	.64487	1.55071	.66986	1.49284	.69545	1.43792	.72167	1.38568	11
50	.64528	1.54972	.67028	1.49190	.69588	1.43703	.72211	1.38484	10
51	.64569	1.54873	.67071	1.49097	.69631	1.43614	.72255	1.38399	9
52	.64610	1.54774	.67113	1.49003	.69675	1.43525	.72299	1.38314	8
53	.64652	1.54675	.67155	1.48909	.69718	1.43436	.72344	1.38229	7
54	.64693	1.54576	.67197	1.48816	.69761	1.43347	.72388	1.38145	6
55	.64734	1.54478	.67239	1.48722	.69804	1.43258	.72432	1.38060	5
56	.64775	1.54379	.67282	1.48629	.69847	1.43169	.72477	1.37976	4
57	.64817	1.54281	.67324	1.48536	.69891	1.43080	.72521	1.37891	3
58	.64858	1.54183	.67366	1.48442	.69934	1.42992	.72565	1.37807	2
59	.64899	1.54085	.67409	1.48349	.69977	1.42903	.72610	1.37722	1
60	.64941	1.53986	.67451	1.48256	.70021	1.42815	.72654	1.37638	0
′	57° Cotang	57° Tang	56° Cotang	56° Tang	55° Cotang	55° Tang	54° Cotang	54° Tang	′

TABLE VII.—*Continued.*

NATURAL TANGENTS AND COTANGENTS.

′	36°		37°		38°		39°		′
	Tang	Cotang	Tang	Cotang	Tang	Cotang	Tang	Cotang	
0	.72654	1.37638	.75355	1.32704	.78129	1.27994	.80978	1.23490	60
1	.72699	1.37554	.75401	1.32624	.78175	1.27917	.81027	1.23416	59
2	.72743	1.37470	.75447	1.32544	.78222	1.27841	.81075	1.23343	58
3	.72788	1.37386	.75492	1.32464	.78269	1.27764	.81123	1.23270	57
4	.72832	1.37302	.75538	1.32384	.78316	1.27688	.81171	1.23196	56
5	.72877	1.37218	.75584	1.32304	.78363	1.27611	.81220	1.23123	55
6	.72921	1.37134	.75629	1.32224	.78410	1.27535	.81268	1.23050	54
7	.72966	1.37050	.75675	1.32144	.78457	1.27458	.81316	1.22977	53
8	.73010	1.36967	.75721	1.32064	.78504	1.27382	.81364	1.22904	52
9	.73055	1.36883	.75767	1.31984	.78551	1.27306	.81413	1.22831	51
10	.73100	1.36800	.75812	1.31904	.78598	1.27230	.81461	1.22758	50
11	.73144	1.36716	.75858	1.31825	.78645	1.27153	.81510	1.22685	49
12	.73189	1.36633	.75904	1.31745	.78692	1.27077	.81558	1.22612	48
13	.73234	1.36549	.75950	1.31666	.78739	1.27001	.81606	1.22539	47
14	.73278	1.36466	.75996	1.31586	.78786	1.26925	.81655	1.22467	46
15	.73323	1.36383	.76042	1.31507	.78834	1.26849	.81703	1.22394	45
16	.73368	1.36300	.76088	1.31427	.78881	1.26774	.81752	1.22321	44
17	.73413	1.36217	.76134	1.31348	.78928	1.26698	.81800	1.22249	43
18	.73457	1.36134	.76180	1.31269	.78975	1.26622	.81849	1.22176	42
19	.73502	1.36051	.76226	1.31190	.79022	1.26546	.81898	1.22104	41
20	.73547	1.35968	.76272	1.31110	.79070	1.26471	.81946	1.22031	40
21	.73592	1.35885	.76318	1.31031	.79117	1.26395	.81995	1.21959	39
22	.73637	1.35802	.76364	1.30952	.79164	1.26319	.82044	1.21886	38
23	.73681	1.35719	.76410	1.30873	.79212	1.26244	.82092	1.21814	37
24	.73726	1.35637	.76456	1.30795	.79259	1.26169	.82141	1.21742	36
25	.73771	1.35554	.76502	1.30716	.79306	1.26093	.82190	1.21670	35
26	.73816	1.35472	.76548	1.30637	.79354	1.26018	.82238	1.21598	34
27	.73861	1.35389	.76594	1.30558	.79401	1.25943	.82287	1.21526	33
28	.73906	1.35307	.76640	1.30480	.79449	1.25867	.82336	1.21454	32
29	.73951	1.35224	.76686	1.30401	.79496	1.25792	.82385	1.21382	31
30	.73996	1.35142	.76733	1.30323	.79544	1.25717	.82434	1.21310	30
31	.74041	1.35060	.76779	1.30244	.79591	1.25642	.82483	1.21238	29
32	.74086	1.34978	.76825	1.30166	.79639	1.25567	.82531	1.21166	28
33	.74131	1.34896	.76871	1.30087	.79686	1.25492	.82580	1.21094	27
34	.74176	1.34814	.76918	1.30009	.79734	1.25417	.82629	1.21023	26
35	.74221	1.34732	.76964	1.29931	.79781	1.25343	.82678	1.20951	25
36	.74267	1.34650	.77010	1.29853	.79829	1.25268	.82727	1.20879	24
37	.74312	1.34568	.77057	1.29775	.79877	1.25193	.82776	1.20808	23
38	.74357	1.34487	.77103	1.29696	.79924	1.25118	.82825	1.20736	22
39	.74402	1.34405	.77149	1.29618	.79972	1.25044	.82874	1.20665	21
40	.74447	1.34323	.77196	1.29541	.80020	1.24969	.82923	1.20593	20
41	.74492	1.34242	.77242	1.29463	.80067	1.24895	.82972	1.20522	19
42	.74538	1.34160	.77289	1.29385	.80115	1.24820	.83022	1.20451	18
43	.74583	1.34079	.77335	1.29307	.80163	1.24746	.83071	1.20379	17
44	.74628	1.33998	.77382	1.29229	.80211	1.24672	.83120	1.20308	16
45	.74674	1.33916	.77428	1.29152	.80258	1.24597	.83169	1.20237	15
46	.74719	1.33835	.77475	1.29074	.80306	1.24523	.83218	1.20166	14
47	.74764	1.33754	.77521	1.28997	.80354	1.24449	.83268	1.20095	13
48	.74810	1.33673	.77568	1.28919	.80402	1.24375	.83317	1.20024	12
49	.74855	1.33592	.77615	1.28842	.80450	1.24301	.83366	1.19953	11
50	.74900	1.33511	.77661	1.28764	.80498	1.24227	.83415	1.19882	10
51	.74946	1.33430	.77708	1.28687	.80546	1.24153	.83465	1.19811	9
52	.74991	1.33349	.77754	1.28610	.80594	1.24079	.83514	1.19740	8
53	.75037	1.33268	.77801	1.28533	.80642	1.24005	.83564	1.19669	7
54	.75082	1.33187	.77848	1.28456	.80690	1.23931	.83613	1.19599	6
55	.75128	1.33107	.77895	1.28379	.80738	1.23858	.83662	1.19528	5
56	.75173	1.33026	.77941	1.28302	.80786	1.23784	.83712	1.19457	4
57	.75219	1.32946	.77988	1.28225	.80834	1.23710	.83761	1.19387	3
58	.75264	1.32865	.78035	1.28148	.80882	1.23637	.83811	1.19316	2
59	.75310	1.32785	.78082	1.28071	.80930	1.23563	.83860	1.19246	1
60	.75355	1.32704	.78129	1.27994	.80978	1.23490	.83910	1.19175	0
′	Cotang	Tang	Cotang	Tang	Cotang	Tang	Cotang	Tang	′
	53°		52°		51°		50°		

TABLE VII.—*Continued.*

NATURAL TANGENTS AND COTANGENTS.

′	40° Tang	40° Cotang	41° Tang	41° Cotang	42° Tang	42° Cotang	43° Tang	43° Cotang	′
0	.83910	1.19175	.86929	1.15037	.90040	1.11061	.93252	1.07237	60
1	.83960	1.19105	.86980	1.14969	.90093	1.10996	.93306	1.07174	59
2	.84009	1.19035	.87031	1.14902	.90146	1.10931	.93360	1.07112	58
3	.84059	1.18964	.87082	1.14834	.90199	1.10867	.93415	1.07049	57
4	.84108	1.18894	.87133	1.14767	.90251	1.10802	.93469	1.06987	56
5	.84158	1.18824	.87184	1.14699	.90304	1.10737	.93524	1.06925	55
6	.84208	1.18754	.87236	1.14632	.90357	1.10672	.93578	1.06862	54
7	.84258	1.18684	.87287	1.14565	.90410	1.10607	.93633	1.06800	53
8	.84307	1.18614	.87338	1.14498	.90463	1.10543	.93688	1.06738	52
9	.84357	1.18544	.87389	1.14430	.90516	1.10478	.93742	1.06676	51
10	.84407	1.18474	.87441	1.14363	.90569	1.10414	.93797	1.06613	50
11	.84457	1.18404	.87492	1.14296	.90621	1.10349	.93852	1.06551	49
12	.84507	1.18334	.87543	1.14229	.90674	1.10285	.93906	1.06489	48
13	.84556	1.18264	.87595	1.14162	.90727	1.10220	.93961	1.06427	47
14	.84606	1.18194	.87646	1.14095	.90781	1.10156	.94016	1.06365	46
15	.84656	1.18125	.87698	1.14028	.90834	1.10091	.94071	1.06303	45
16	.84706	1.18055	.87749	1.13961	.90887	1.10027	.94125	1.06241	44
17	.84756	1.17986	.87801	1.13894	.90940	1.09963	.94180	1.06179	43
18	.84806	1.17916	.87852	1.13828	.90993	1.09899	.94235	1.06117	42
19	.84856	1.17846	.87904	1.13761	.91046	1.09834	.94290	1.06056	41
20	.84906	1.17777	.87955	1.13694	.91099	1.09770	.94345	1.05994	40
21	.84956	1.17708	.88007	1.13627	.91153	1.09706	.94400	1.05932	39
22	.85006	1.17638	.88059	1.13561	.91206	1.09642	.94455	1.05870	38
23	.85057	1.17569	.88110	1.13494	.91259	1.09578	.94510	1.05809	37
24	.85107	1.17500	.88162	1.13428	.91313	1.09514	.94565	1.05747	36
25	.85157	1.17430	.88214	1.13361	.91366	1.09450	.94620	1.05685	35
26	.85207	1.17361	.88265	1.13295	.91419	1.09386	.94676	1.05624	34
27	.85257	1.17292	.88317	1.13228	.91473	1.09322	.94731	1.05562	33
28	.85308	1.17223	.88369	1.13162	.91526	1.09258	.94786	1.05501	32
29	.85358	1.17154	.88421	1.13096	.91580	1.09195	.94841	1.05439	31
30	.85408	1.17085	.88473	1.13029	.91633	1.09131	.94896	1.05378	30
31	.85458	1.17016	.88524	1.12963	.91687	1.09067	.94952	1.05317	29
32	.85509	1.16947	.88576	1.12897	.91740	1.09003	.95007	1.05255	28
33	.85559	1.16878	.88628	1.12831	.91794	1.08940	.95062	1.05194	27
34	.85609	1.16809	.88680	1.12765	.91847	1.08876	.95118	1.05133	26
35	.85660	1.16741	.88732	1.12699	.91901	1.08813	.95173	1.05072	25
36	.85710	1.16672	.88784	1.12633	.91955	1.08749	.95229	1.05010	24
37	.85761	1.16603	.88836	1.12567	.92008	1.08686	.95284	1.04949	23
38	.85811	1.16535	.88888	1.12501	.92062	1.08622	.95340	1.04888	22
39	.85862	1.16466	.88940	1.12435	.92116	1.08559	.95395	1.04827	21
40	.85912	1.16398	.88992	1.12369	.92170	1.08496	.95451	1.04766	20
41	.85963	1.16329	.89045	1.12303	.92224	1.08432	.95506	1.04705	19
42	.86014	1.16261	.89097	1.12238	.92277	1.08369	.95562	1.04644	18
43	.86064	1.16192	.89149	1.12172	.92331	1.08306	.95618	1.04583	17
44	.86115	1.16124	.89201	1.12106	.92385	1.08243	.95673	1.04522	16
45	.86166	1.16056	.89253	1.12041	.92439	1.08179	.95729	1.04461	15
46	.86216	1.15987	.89306	1.11975	.92493	1.08116	.95785	1.04401	14
47	.86267	1.15919	.89358	1.11909	.92547	1.08053	.95841	1.04340	13
48	.86318	1.15851	.89410	1.11844	.92601	1.07990	.95897	1.04279	12
49	.86368	1.15783	.89463	1.11778	.92655	1.07927	.95952	1.04218	11
50	.86419	1.15715	.89515	1.11713	.92709	1.07864	.96008	1.04158	10
51	.86470	1.15647	.89567	1.11648	.92763	1.07801	.96064	1.04097	9
52	.86521	1.15579	.89620	1.11582	.92817	1.07738	.96120	1.04036	8
53	.86572	1.15511	.89672	1.11517	.92872	1.07676	.96176	1.03976	7
54	.86623	1.15443	.89725	1.11452	.92926	1.07613	.96232	1.03915	6
55	.86674	1.15375	.89777	1.11387	.92980	1.07550	.96288	1.03855	5
56	.86725	1.15308	.89830	1.11321	.93034	1.07487	.96344	1.03794	4
57	.86776	1.15240	.89883	1.11256	.93088	1.07425	.96400	1.03734	3
58	.86827	1.15172	.89935	1.11191	.93143	1.07362	.96457	1.03674	2
59	.86878	1.15104	.89988	1.11126	.93197	1.07299	.96513	1.03613	1
60	.86929	1.15037	.90040	1.11061	.93252	1.07237	.96569	1.03553	0
′	Cotang	Tang	Cotang	Tang	Cotang	Tang	Cotang	Tang	′
	49°		48°		47°		46°		

TABLE VII.—*Continued.*

Natural Tangents and Cotangents.

′	44°		′	′	44°		′	′	44°		′
	Tang	Cotang			Tang	Cotang			Tang	Cotang	
0	.96569	1.03553	60	20	.97700	1.02355	40	40	.98843	1.01170	20
1	.96625	1.03493	59	21	.97756	1.02295	39	41	.98901	1.01112	19
2	.96681	1.03433	58	22	.97813	1.02236	38	42	.98958	1 01053	18
3	.96738	1.03372	57	23	.97870	1.02176	37	43	.99016	1.00994	17
4	.96794	1.03312	56	24	.97927	1.02117	36	44	.99073	1.00935	16
5	.96850	1.03252	55	25	.97984	1.02057	35	45	.99131	1.00876	15
6	.96907	1.03192	54	26	.98041	1.01998	34	46	.99189	1.00818	14
7	.96963	1.03132	53	27	.98098	1.01939	33	47	.99247	1.00759	13
8	.97020	1.03072	52	28	.98155	1.01879	32	48	.99304	1.00701	12
9	.97076	1.03012	51	29	.98213	1.01820	31	49	.99362	1.00642	11
10	.97133	1.02952	50	30	.98270	1.01761	30	50	.99420	1.00583	10
11	.97189	1.02892	49	31	.98327	1.01702	29	51	.99478	1.00525	9
12	.97246	1.02832	48	32	.98384	1.01642	28	52	.99536	1.00467	8
13	.97302	1.02772	47	33	.98441	1.01583	27	53	.99594	1.00408	7
14	.97359	1.02713	46	34	.98499	1.01524	26	54	.99652	1.00350	6
15	.97416	1.02653	45	35	.98556	1.01465	25	55	.99710	1.00291	5
16	.97472	1.02593	44	36	.98613	1.01406	24	56	.99768	1.00233	4
17	.97529	1.02533	43	37	.98671	1.01347	23	57	.99826	1.00175	3
18	.97586	1.02474	42	38	.98728	1.01288	22	58	.99884	1.00116	2
19	.97643	1.02414	41	39	.98786	1.01229	21	59	.99942	1.00058	1
20	.97700	1.02355	40	40	.98843	1.01170	20	60	1.00000	1.00000	0
′	Cotang	Tang	′	′	Cotang	Tang	′	′	Cotang	Tang	′
	45°				45°				45°		

TABLE VIII.

CO-ORDINATES OF POINTS OF INTERSECTION OF PARALLELS AND MERIDIANS IN POLYCONIC PROJECTION. § 417.

Latitude.	Length of 1° Latitude, in Statute Miles.	Length of Side of Tangent Cone, in Statute Miles.	Meridian Distances for 1° Longitude.				Divergence of Parallels for 1° Longitude.			
			In Yards.	In Metres.	In Miles.	Factor.	In Yards.	In Metres.	In Miles.	Factor.
30°	68.875	6869	105507	96476	59.95	$n \cos (0.288n^\circ)$	460.4	421.0	0.2617	n^2
32°	68.897	6348	103327	94481	58.71	$n \cos (0.304n^\circ)$	477.8	436.8	0.2715	n^2
34°	68.918	5881	101022	92373	57.40	$n \cos (0.320n^\circ)$	493.0	450.7	0.2800	n^2
36°	68.941	5461	98593	90152	56.02	$n \cos (0.337n^\circ)$	505.7	462.4	0.2873	n^2
38°	68.964	5079	96044	87822	54.57	$n \cos (0.353n^\circ)$	516.0	471.8	0.2932	n^2
40°	68.987	4729	93377	85383	53.06	$n \cos (0.369n^\circ)$	523.8	479.0	0.2976	n^2
42°	69.011	4408	90596	82840	51.48	$n \cos (0.386n^\circ)$	529.0	483.8	0.3006	n^2
44°	69.036	4110	87704	80197	49.83	$n \cos (0.402n^\circ)$	531.7	486.2	0.3022	n^2
46°	69.060	3833	84704	77452	48.13	$n \cos (0.418n^\circ)$	531.7	486.2	0.3022	n^2
48°	69.084	3575	81601	74615	46.37	$n \cos (0.435n^\circ)$	529.2	484.0	0.3007	n^2
50°	69.108	3332	78398	71686	44.54	$n \cos (0.451n^\circ)$	524.1	479.2	0.2978	n^2

n = number degrees of longitude between the given meridian and the prime meridian of the map.

TABLE IX.

GIVING VALUES OF C IN KUTTER'S FORMULA WHEN $s = 0.001$. § 259.

r in feet.	Values of n. .009	.010	.011	.012	.013	.015	.017	.020	.0225	.025	.030	.035	r in feet.
.1	108.3	93.8	82.2	72.7	65.0	53.2	44.6	35.5	30.0	25.9	20.1	16.3	.1
.2	129.5	113.1	100.0	89.1	80.2	66.3	56.2	45.2	38.6	33.6	26.3	21.5	.2
.3	141.8	123.8	111.0	98.8	90.2	75.0	63.4	51.8	44.6	38.4	30.3	25.1	.3
.4	150.2	132.5	118.0	106.0	96.2	80.4	68.8	56.2	48.4	42.4	33.7	27.8	.4
.5	156.8	138.6	123.8	111.2	101.2	85.1	72.8	60.0	51.8	45.4	36.2	30.0	.5
.6	161.9	143.3	128.3	115.7	105.3	88.8	76.4	62.9	54.5	48.0	38.5	32.0	.6
.7	166.1	147.4	131.9	119.3	108.7	92.0	79.3	65.4	56.9	50.2	40.3	33.6	.7
.8	169.7	150.8	135.1	122.3	111.6	94.6	81.9	67.7	59.0	52.2	42.0	35.1	.8
.9	172.8	153.7	137.8	125.1	114.2	97.0	84.2	69.7	60.8	53.8	43.4	36.3	.9
1.0	175.4	156.2	140.5	127.4	116.5	99.1	86.0	71.5	62.5	55.4	44.9	37.7	1.0
1.2	180.0	160.4	144.6	131.5	120.4	102.7	89.4	74.5	65.3	58.1	47.1	39.7	1.2
1.4	183.6	164.0	147.9	134.7	123.7	105.7	92.2	77.0	67.7	60.2	49.2	41.5	1.4
1.6	186.7	167.0	150.8	137.4	126.2	108.2	94.5	79.3	69.9	62.3	51.0	43.2	1.6
1.8	189.2	169.5	153.2	139.7	128.7	110.3	96.6	81.1	71.6	64.0	52.6	44.6	1.8
2.0	191.4	171.6	155.4	141.8	130.5	112.3	98.4	82.9	73.4	65.5	54.0	45.9	2.0
2.2	193.3	173.5	157.3	143.7	132.3	114.0	100.0	84.3	74.7	66.9	55.2	47.0	2.2
2.4	195.0	175.2	159.0	145.3	133.9	115.4	101.4	85.6	76.0	68.1	56.3	48.0	2.4
2.6	196.7	176.8	160.5	146.8	135.3	116.8	102.8	87.0	77.1	69.2	57.4	49.0	2.6
2.8	198.0	178.2	161.8	148.1	136.7	118.0	104.0	88.2	78.2	70.3	58.4	49.9	2.8
3.0	199.3	179.4	163.2	149.3	137.9	119.2	105.1	89.3	79.2	71.3	59.2	50.6	3.0
3.4	201.7	181.7	165.3	151.4	140.0	121.3	107.1	91.1	81.0	73.0	60.8	52.0	3.4
3.8	203.7	183.6	167.2	153.3	141.8	123.0	108.8	92.7	82.5	74.5	62.3	53.5	3.8
4.2	205.4	185.3	168.8	155.0	143.4	124.6	110.3	94.3	83.9	75.8	63.5	54.7	4.2
4.6	207.0	186.8	170.3	156.4	144.8	125.9	111.6	95.4	85.2	77.0	64.7	55.7	4.6
5.0	208.3	188.1	171.6	157.7	146.0	127.2	112.9	96.6	86.3	78.1	65.7	56.7	5.0

TABLE X. § 259.

GIVING DIAMETERS IN FEET OF CIRCULAR BRICK CONDUITS FOR VARIOUS INCLINATIONS AND RATES OF DISCHARGE.

Conduit full to point of maximum discharge. (By Kutter's formula.) § 259.

Cubic Feet per Second.	Fall per 100 Feet.													Cubic Feet per Second.
	.10	.15	.20	.30	.50	.75	1.0	1.5	2.0	3.0	4.0	6.0	10.0	
1	.9	.9	.8	.8	.7	.7	.6	.6	.6	.5	.5	.5	.4	1
2	1.2	1.1	1.1	1.0	.9	.8	.8	.7	.7	.7	.6	.6	.5	2
3	1.4	1.3	1.3	1.2	1.1	1.0	.9	.9	.8	.8	.7	.7	.6	3
4	1.6	1.5	1.4	1.3	1.2	1.1	1.0	1.0	.9	.8	.8	.7	.7	4
5	1.7	1.6	1.5	1.4	1.3	1.2	1.1	1.0	1.0	.9	.9	.8	.7	5
10	2.2	2.0	1.9	1.8	1.6	1.5	1.4	1.3	1.3	1.2	1.1	1.0	.9	10
15	2.6	2.4	2.3	2.1	1.9	1.8	1.7	1.6	1.5	1.4	1.3	1.2	1.1	15
20	2.9	2.6	2.5	2.3	2.1	2.0	1.9	1.7	1.6	1.5	1.4	1.3	1.2	20
25	3.1	2.9	2.7	2.5	2.3	2.1	2.0	1.9	1.8	1.6	1.6	1.5	1.3	25
30	3.3	3.1	2.9	2.7	2.5	2.3	2.2	2.0	1.9	1.8	1.7	1.5	1.4	30
35	3.5	3.3	3.1	2.9	2.6	2.4	2.3	2.1	2.0	1.9	1.8	1.6	1.5	35
40	3.7	3.4	3.2	3.0	2.7	2.5	2.4	2.2	2.1	2.0	1.9	1.7	1.6	40
45	3.9	3.6	3.4	3.1	2.9	2.6	2.5	2.3	2.2	2.0	1.9	1.8	1.6	45
50	4.0	3.7	3.5	3.3	3.0	2.7	2.6	2.4	2.3	2.1	2.0	1.9	1.7	50
60	4.3	4.0	3.7	3.5	3.2	2.9	2.8	2.6	2.5	2.3	2.1	2.0	1.8	60
70	4.5	4.2	4.0	3.7	3.4	3.1	2.9	2.7	2.6	2.4	2.3	2.1	1.9	70
80	4.8	4.4	4.2	4.0	3.5	3.3	3.1	2.9	2.7	2.5	2.4	2.2	2.0	80
90	5.0	4.6	4.4	4.1	3.7	3.4	3.2	3.0	2.9	2.7	2.5	2.3	2.1	90
100	5.2	4.8	4.6	4.2	3.8	3.6	3.4	3.1	3.0	2.7	2.6	2.4	2.2	100
125	5.7	5.2	5.0	4.6	4.2	3.9	3.6	3.4	3.2	3.0	2.8	2.6	2.4	125
150	6.1	5.6	5.3	4.9	4.5	4.1	3.9	3.6	3.4	3.2	3.0	2.8	2.6	150
175	6.4	5.9	5.6	5.2	4.6	4.4	4.2	3.8	3.6	3.4	3.2	3.0	2.7	175
200	6.8	6.3	5.9	5.5	5.0	4.6	4.4	4.0	3.8	3.6	3.4	3.1	2.8	200
250	7.3	6.8	6.4	5.9	5.4	5.0	4.7	4.4	4.2	3.9	3.7	3.4	3.1	250
300	7.9	7.3	6.9	6.4	5.8	5.4	5.1	4.7	4.5	4.1	3.9	3.6	3.3	300
350	8.3	7.7	7.3	6.8	6.1	5.7	5.4	5.0	4.7	4.4	4.2	3.9	3.5	350
400	8.8	8.1	7.7	7.1	6.5	6.0	5.7	5.2	5.0	4.6	4.4	4.0	3.7	400
500	9.5	8.8	8.4	7.8	7.0	6.5	6.2	5.7	5.4	5.0	4.7	4.4	4.0	500
600	10.2	9.5	9.0	8.3	7.5	7.0	6.6	6.1	5.8	5.4	5.1	4.7	4.3	600
700	10.8	10.0	9.5	8.8	8.0	7.4	7.0	6.5	6.1	5.7	5.4	5.0	4.5	700
800	11.4	10.6	10.0	9.3	8.4	7.8	7.4	6.8	6.5	6.0	5.7	5.3	4.7	800
900	11.9	11.1	10.5	9.7	8.8	8.1	7.7	7.1	6.7	6.2	5.9	5.5	5.0	900
1,000	12.4	11.5	10.9	10.1	9.2	8.5	8.0	7.4	7.0	6.5	6.2	5.7	5.2	1,000

TABLE XI.

VOLUMES BY THE PRISMOIDAL FORMULA. § 320.

Widths.	Heights. 1	2	3	4	5	6	7	8	9	10	Corrections for tenths in height.	
1	0	1	1	1	2	2	2	2	3	3	.1	0
2	1	1	2	2	3	3	4	5	6	6	.2	0
3	1	2	3	4	5	6	6	7	8	9	.3	0
4	1	2	4	5	6	7	9	10	11	12	.4	1
5	−2	−3	−5	−6	−8	−9	−11	−12	−14	−15	.5	1
6	2	4	6	7	9	11	13	15	17	19	.6	1
7	2	4	6	9	11	13	15	17	19	22	.7	1
8	2	5	7	10	12	15	17	20	22	25	.8	1
9	3	6	8	11	14	17	19	22	25	28	.9	1
10	3	6	9	12	15	19	22	25	28	31		
11	3	7	10	14	17	20	24	27	31	34	.1	0
12	4	7	11	15	19	22	26	30	33	37	.2	1
13	4	8	12	16	20	24	28	32	36	40	.3	1
14	4	9	13	17	22	26	30	35	39	43	.4	2
15	−5	−9	−14	−19	−23	−28	−32	−37	−42	−46	.5	2
16	5	10	15	20	25	30	35	40	44	49	.6	3
17	5	10	16	21	26	31	37	42	47	52	.7	3
18	6	11	17	22	28	33	39	44	50	56	.8	4
19	6	12	18	23	29	35	41	47	53	59	.9	4
20	6	12	19	25	31	37	43	49	56	62		
21	6	13	19	26	32	39	45	52	58	65	.1	1
22	7	14	20	27	34	41	48	54	61	68	.2	2
23	7	14	21	28	35	43	50	57	64	71	.3	2
24	7	15	22	30	37	44	52	59	67	74	.4	3
25	−8	−15	−23	−31	−39	−46	−54	−62	−69	−77	.5	4
26	8	16	24	32	40	48	56	64	72	80	.6	5
27	8	17	25	33	42	50	58	67	75	83	.7	5
28	9	17	26	35	43	52	60	69	78	86	.8	6
29	9	18	27	36	45	54	63	72	81	90	.9	7
30	9	19	28	37	46	56	65	74	83	93		
31	10	19	29	38	48	57	67	77	86	96	.1	1
32	10	20	30	40	49	59	69	79	89	99	.2	2
33	10	20	31	41	51	61	71	81	92	102	.3	3
34	10	21	31	42	52	63	73	84	94	105	.4	4
35	−11	−22	−32	−43	−54	−65	−76	−86	−97	−108	.5	5
36	11	22	33	44	56	67	78	89	100	111	.6	6
37	11	23	34	46	57	69	80	91	103	114	.7	8
38	12	23	35	47	59	70	82	94	106	117	.8	9
39	12	24	36	48	60	72	84	96	108	120	.9	10
40	12	25	37	49	62	74	86	99	111	123		
41	13	25	38	51	63	76	89	101	114	127	.1	1
42	13	26	39	52	65	78	91	104	117	130	.2	3
43	13	27	40	53	66	80	93	106	119	133	.3	4
44	14	27	41	54	68	81	95	109	122	136	.4	6
45	−14	−28	−42	−56	−69	−83	−97	−111	−125	−139	.5	7
46	14	28	43	57	71	85	99	114	128	142	.6	8
47	15	29	44	58	73	87	102	116	131	145	.7	10
48	15	30	44	59	74	89	104	119	133	148	.8	11
49	15	30	45	60	76	91	106	121	136	151	.9	13
50	15	31	46	62	77	93	108	123	139	154		
	1	2	3	4	5	6	7	8	9	10		
	.1	.2	.3	.4	.5	.6	.7	.8	.9	Corrections for tenths in width.		
	0	0	0	1	1	1	1	1	1			

TABLE XI.—*Continued.*

Volumes by the Prismoidal Formula.

Widths.	Heights. 1	2	3	4	5	6	7	8	9	10	Corrections for tenths in height.	
51	16	31	47	63	79	94	110	126	142	157	.1	2
52	16	32	48	64	80	96	112	128	144	160	.2	3
53	16	33	49	65	82	98	115	131	147	163	.3	5
54	17	33	50	67	83	100	117	133	150	167	.4	7
55	—17	—34	—51	—68	—85	—102	—119	—136	—153	—170	.5	8
56	17	35	52	69	86	104	121	138	156	173	.6	10
57	18	35	53	70	88	106	123	141	158	•176	.7	12
58	18	36	54	72	90	107	125	143	161	179	.8	14
59	18	36	55	73	91	109	127	146	164	182	.9	15
60	19	37	56	74	93	111	130	148	167	185		
61	19	38	56	75	94	113	132	151	169	188	.1	2
62	19	38	57	77	96	115	134	153	172	191	.2	4
63	19	39	58	78	97	117	136	156	175	194	.3	6
64	20	40	59	79	99	119	138	158	178	197	.4	8
65	—20	—40	—60	—80	—100	—120	—140	—160	—181	—201	.5	10
66	20	41	61	81	102	122	143	163	183	204	.6	12
67	21	41	62	83	103	124	145	165	186	207	.7	14
68	21	42	63	84	105	126	147	168	189	210	.8	16
69	21	43	64	85	106	128	149	170	192	213	.9	18
70	22	43	65	86	108	130	151	173	194	216		
71	22	44	66	88	100	131	153	175	197	219	.1	2
72	22	44	67	89	111	133	156	178	200	222	.2	5
73	23	45	68	90	113	135	158	180	203	225	.3	7
74	23	46	69	91	114	137	160	183	206	228	.4	9
75	—23	—46	—69	—93	—116	—139	—162	—185	—208	—231	.5	12
76	23	47	70	94	117	141	164	188	211	235	.6	14
77	24	48	71	95	119	143	166	190	214	238	.7	16
78	24	48	72	96	120	144	169	193	217	241	.8	19
79	24	49	73	98	122	146	171	195	219	244	.9	21
80	25	49	74	99	123	148	173	198	222	247		
81	25	50	75	100	125	150	175	200	225	250	.1	3
82	25	51	76	101	127	152	177	202	228	253	.2	5
83	26	51	77	102	128	154	179	205	231	256	.3	8
84	26	52	78	104	130	156	181	207	233	259	.4	10
85	—26	—52	—79	—105	—131	—157	—184	—210	—236	—262	.5	13
86	27	53	80	106	133	159	186	212	239	265	.6	16
87	27	54	81	107	134	161	188	215	242	269	.7	18
88	27	54	81	109	136	163	190	217	244	272	.8	21
89	27	55	82	110	137	165	192	220	247	275	.9	24
90	28	56	83	111	139	167	194	222	250	278		
91	28	56	84	112	140	169	197	225	253	281	.1	3
92	28	57	85	114	142	170	199	227	256	284	.2	6
93	29	57	86	115	144	172	201	230	258	287	.3	9
94	29	58	87	116	145	174	203	232	261	290	.4	12
95	—29	—59	—88	—117	—147	—176	—205	—235	—264	—293	.5	15
96	30	59	89	119	148	178	207	237	267	296	.6	18
97	30	60	90	120	150	180	210	240	269	299	.7	21
98	30	60	91	121	151	181	212	242	272	302	.8	23
99	31	61	92	122	153	183	214	244	275	306	.9	26
100	31	62	93	123	154	185	216	247	278	309		
	1	2	3	4	5	6	7	8	9	10		
	.1	.2	.3	.4	.5	.6	.7	.8	.9	Corrections for tenths in width.		
	0	0	0	1	1	1	1	1	1			

TABLE XI.—*Continued.*

VOLUMES BY THE PRISMOIDAL FORMULA.

Widths.	Heights. 11	12	13	14	15	16	17	18	19	20	Corrections for tenths in height.	
1	3	4	4	4	5	5	5	6	6	6	.1	0
2	7	7	8	9	9	10	10	11	12	12	.2	0
3	10	11	12	13	14	15	16	17	18	19	.3	0
4	14	15	16	17	19	20	21	22	23	25	.4	1
5	—17	—19	—20	—22	—23	—25	—26	—28	—29	—31	.5	1
6	20	22	24	26	28	30	31	33	35	37	.6	1
7	24	26	28	30	32	35	37	39	41	43	.7	1
8	27	30	32	35	37	40	42	44	47	49	.8	1
9	31	33	36	39	42	44	47	50	53	56	.9	1
10	34	37	40	43	46	49	52	56	59	62		
11	37	41	44	48	51	54	58	61	65	68	.1	0
12	41	44	48	52	56	59	63	67	70	74	.2	1
13	44	48	52	56	60	64	68	72	76	80	.3	1
14	48	52	56	60	65	69	73	78	82	86	.4	2
15	—51	—56	—60	—65	—69	—74	—79	—83	—88	—93	.5	2
16	54	59	64	69	74	79	84	89	94	99	.6	3
17	58	63	68	73	79	84	89	94	100	105	.7	3
18	61	67	72	78	83	89	94	100	106	111	.8	4
19	65	70	76	82	88	94	100	106	111	117	.9	4
20	68	74	80	86	93	99	105	111	117	123		
21	71	78	84	91	97	104	110	117	123	130	.1	1
22	75	81	88	95	102	109	115	122	129	136	.2	2
23	78	85	92	99	106	114	121	128	135	142	.3	2
24	81	89	96	104	111	119	126	133	141	148	.4	3
25	—85	—93	—100	—108	—116	—123	—131	—139	—147	—154	.5	4
26	88	96	104	112	120	128	136	144	152	160	.6	5
27	92	100	108	117	125	133	142	150	158	167	.7	5
28	95	104	112	121	130	138	147	156	164	173	.8	6
29	98	107	116	125	134	143	152	161	170	179	.9	7
30	102	111	120	130	139	148	157	167	176	185		
31	105	115	124	134	144	153	163	172	182	191	.1	1
32	109	119	128	138	148	158	168	178	188	198	.2	2
33	112	122	132	143	153	163	173	183	194	204	.3	3
34	115	126	136	147	157	168	178	189	199	210	.4	4
35	—119	—130	—140	—151	—162	—173	—184	—194	—205	—216	.5	5
36	122	133	144	156	167	178	189	200	211	222	.6	6
37	126	137	148	160	171	183	194	206	217	228	.7	8
38	129	141	152	164	176	188	199	211	223	235	.8	9
39	132	144	156	169	181	193	205	217	229	241	.9	10
40	136	148	160	173	185	198	210	222	235	247		
41	139	152	165	177	190	202	215	228	240	253	.1	1
42	143	156	169	181	194	207	220	233	246	259	.2	3
43	146	159	173	186	199	212	226	239	252	265	.3	4
44	149	163	177	190	204	217	231	244	258	272	.4	6
45	—153	—167	—181	—194	—208	—222	—236	—250	—264	—278	.5	7
46	156	170	183	199	213	227	241	256	270	284	.6	8
47	160	174	189	203	218	232	247	261	276	290	.7	10
48	163	178	193	207	222	237	252	267	281	296	.8	11
49	166	181	197	212	227	242	257	272	287	302	.9	13
50	170	185	201	216	231	247	262	278	293	309		
	11	12	13	14	15	16	17	18	19	20		
	.1	.2	.3	.4	.5	.6	.7	.8	.9	Corrections for tenths in width.		
	0	1	1	2	2	3	3	4	4			

TABLE XI.—*Continued.*

Volumes by the Prismoidal Formula.

Widths.	Heights. 11	12	13	14	15	16	17	18	19	20	Corrections for tenths in height.	
51	173	189	205	220	236	252	268	283	299	315	.1	2
52	177	193	209	225	241	257	273	289	305	321	.2	3
53	180	196	213	229	245	262	278	294	311	327	.3	5
54	183	200	217	233	250	267	283	300	317	333	.4	7
55	—187	—204	—221	—238	—255	—272	—289	—306	—323	—340	.5	8
56	190	207	225	242	259	277	294	311	328	346	.6	10
57	194	211	229	246	264	281	299	317	334	352	.7	12
58	197	215	233	251	269	286	304	322	340	358	.8	14
59	200	219	237	255	273	291	310	328	346	364	.9	15
60	204	222	241	259	278	296	315	333	352	370		
61	207	226	245	264	282	301	320	339	358	377	.1	2
62	210	230	249	268	287	306	325	344	364	383	.2	4
63	214	233	253	272	292	311	331	350	369	389	.3	6
64	217	237	257	277	296	316	336	356	375	395	.4	8
65	—221	—241	—261	—281	—301	—321	—341	—361	—381	—401	.5	10
66	224	244	265	285	306	326	346	367	387	407	.6	12
67	227	248	269	290	310	331	352	372	393	414	.7	14
68	231	252	273	294	315	336	357	378	399	420	.8	16
69	234	256	277	298	319	341	362	383	405	426	.9	18
70	238	259	281	302	324	346	367	389	410	432		
71	241	263	285	307	329	351	373	394	416	438	.1	2
72	244	267	289	311	333	356	378	400	422	444	.2	5
73	248	270	293	315	338	360	383	406	428	451	.3	7
74	251	274	297	320	343	365	388	411	434	457	.4	9
75	—255	—278	—301	—324	—347	—370	—394	—417	—440	—463	.5	12
76	258	281	305	328	352	375	399	422	446	469	.6	14
77	261	285	309	333	356	380	40[illegible]	428	452	475	.7	16
78	265	289	313	337	361	385	409	433	457	481	.8	19
79	268	293	317	341	366	390	415	439	463	488	.9	21
80	272	296	321	346	370	395	420	444	469	494		
81	275	300	325	350	375	400	425	450	475	500	.1	3
82	278	304	329	354	380	405	430	456	481	506	.2	5
83	282	307	333	359	384	410	435	461	487	512	.3	8
84	285	311	337	363	389	415	441	467	493	519	.4	10
85	—289	—315	—341	—367	—394	—420	—446	—472	—498	—525	.5	13
86	292	319	345	372	398	425	451	478	504	531	.6	16
87	295	322	349	376	403	430	456	483	510	537	.7	18
88	299	326	353	380	407	435	462	489	516	543	.8	21
89	303	330	357	385	412	440	467	494	522	549	.9	24
90	306	333	361	389	417	444	472	500	528	556		
91	309	337	365	393	421	449	477	506	534	562	.1	3
92	312	341	369	398	426	454	483	511	540	568	.2	6
93	316	344	373	402	431	459	488	517	545	574	.3	9
94	319	348	377	406	435	464	493	522	551	580	.4	12
95	—323	—352	—381	—410	—440	—469	—498	—528	—557	—586	.5	15
96	326	356	385	415	444	474	504	533	563	593	.6	18
97	329	359	389	419	449	479	509	539	569	599	.7	21
98	333	363	393	423	454	484	514	544	575	605	.8	23
99	336	367	397	428	458	489	519	550	581	611	.9	26
100	340	370	401	432	463	494	525	556	586	617		
	11	12	13	14	15	16	17	18	19	20		
	.1	.2	.3	.4	.5	.6	.7	.8	.9	Corrections for tenths in width.		
	0	1	1	2	2	3	3	4	4			

TABLE XI.—*Continued.*

VOLUMES BY THE PRISMOIDAL FORMULA.

Widths.	Heights. 21	22	23	24	25	26	27	28	29	30	Corrections for tenths in height.	
1	6	7	7	7	8	8	8	9	9	9	.1	0
2	13	14	14	15	15	16	17	17	18	19	.2	0
3	19	20	21	22	23	24	25	26	27	28	.3	0
4	26	27	28	30	31	32	33	35	36	37	.4	1
5	—32	—34	—35	—37	—39	—40	—42	—43	—45	—46	.5	1
6	39	41	43	44	46	48	50	52	54	56	.6	1
7	45	48	50	52	54	56	58	60	63	65	.7	1
8	52	54	57	59	62	64	67	69	72	74	.8	1
9	58	61	64	67	69	72	75	78	81	83	.9	1
10	65	68	71	74	77	80	83	86	90	93		
11	71	75	78	81	85	88	92	95	98	102	.1	0
12	78	81	85	89	93	96	100	104	107	111	.2	1
13	84	88	92	96	100	114	108	112	116	120	.3	1
14	91	95	99	104	108	112	117	121	125	130	.4	2
15	—97	—102	—106	—111	—116	—120	—125	—130	—134	—139	.5	2
16	104	109	114	119	123	128	133	138	143	148	.6	3
17	110	115	121	126	131	136	142	147	152	157	.7	3
18	117	122	128	133	139	144	150	156	161	167	.8	4
19	123	129	135	141	147	152	158	164	170	176	.9	4
20	130	136	142	148	154	160	167	173	179	185		
21	136	142	149	156	162	169	175	181	188	194	.1	1
22	143	149	156	163	170	177	183	190	197	204	.2	2
23	149	156	163	170	177	185	192	199	206	213	.3	2
24	156	163	170	178	185	193	200	207	215	222	.4	3
25	—162	—170	—177	—185	—193	—201	—208	—216	—224	—231	.5	4
26	169	177	185	193	201	209	217	225	233	241	.6	5
27	175	183	192	200	208	217	225	233	242	250	.7	5
28	181	190	199	207	216	225	233	242	251	259	.8	6
29	188	197	206	215	224	233	242	251	260	269	.9	7
30	194	204	213	222	231	241	250	259	269	278		
31	201	210	220	230	239	249	258	268	277	287	.1	1
32	207	217	227	237	247	257	267	277	286	296	.2	2
33	214	224	234	244	255	265	275	285	295	306	.3	3
34	220	231	241	252	262	273	283	294	304	315	.4	4
35	—227	—238	—248	—259	—270	—281	—292	—302	—313	—324	.5	5
36	233	244	256	267	278	289	300	311	322	333	.6	6
37	240	251	263	274	285	297	308	320	331	343	.7	8
38	246	258	270	281	293	305	317	328	340	352	.8	9
39	253	265	277	289	301	313	325	337	349	361	.9	10
40	259	272	284	296	309	321	333	346	358	370		
41	266	278	291	304	316	329	342	354	367	380	.1	1
42	272	285	298	311	324	337	350	363	376	389	.2	3
43	279	292	305	319	332	345	358	372	385	398	.3	4
44	285	299	312	326	340	353	367	380	394	407	.4	6
45	—292	—306	—319	—333	—347	—361	—375	—389	—403	—417	.5	7
46	298	312	327	341	355	369	383	398	412	426	.6	8
47	305	319	334	348	363	377	392	406	421	435	.7	10
48	311	326	341	356	370	385	400	415	430	444	.8	11
49	318	333	348	363	378	393	408	423	439	454	.9	13
50	324	340	355	370	386	401	417	432	448	463		
	21	22	23	24	25	26	27	28	29	30		
	.1	.2	.3	.4	.5	.6	.7	.8	.9	Corrections for tenths in width.		
	1	2	2	3	4	5	5	6	7			

TABLE XI.—*Continued.*

VOLUMES BY THE PRISMOIDAL FORMULA.

Widths.	Heights. 21	22	23	24	25	26	27	28	29	30	Corrections for tenths in height.	
51	331	346	362	378	394	409	425	441	456	472	.1	2
52	337	353	369	385	401	417	433	449	465	481	.2	3
53	344	360	376	393	409	425	442	458	474	491	.3	5
54	350	367	383	400	417	433	450	467	483	500	.4	7
55	—356	—373	—390	—407	—424	—441	—458	—475	—492	—509	.5	8
56	363	380	398	415	432	449	467	484	501	519	.6	10
57	369	387	405	422	440	457	475	493	510	528	.7	12
58	376	394	412	430	448	465	483	501	519	537	.8	14
59	382	401	419	437	455	473	492	510	528	546	.9	15
60	389	407	426	444	463	481	500	519	537	556		
61	395	414	433	452	471	490	508	527	546	565	.1	2
62	402	421	440	459	478	498	517	536	555	574	.2	4
63	408	428	447	467	486	506	525	544	564	583	.3	6
64	415	435	454	474	494	514	533	553	573	593	.4	8
65	—421	—441	—461	—481	—502	—522	—542	—562	—582	—602	.5	10
66	428	448	469	489	509	530	550	570	591	611	.6	12
67	431	455	476	496	517	538	558	579	600	620	.7	14
68	441	462	483	504	525	546	567	588	609	630	.8	16
69	447	469	490	511	532	554	575	596	618	639	.9	18
70	454	475	497	519	540	562	583	605	627	648		
71	460	482	504	526	548	570	592	614	635	657	.1	2
72	467	489	511	533	556	578	600	622	644	667	.2	5
73	473	496	518	541	563	586	608	631	653	676	.3	7
74	480	502	525	548	571	594	617	640	662	685	.4	9
75	—486	—509	—532	—556	—579	—601	—625	—648	—671	—694	.5	12
76	493	516	540	563	586	610	633	657	680	704	.6	14
77	499	523	547	570	594	618	642	665	689	713	.7	16
78	506	530	554	578	602	626	650	674	698	722	.8	19
79	512	536	561	585	610	634	658	683	707	731	.9	21
80	519	543	568	593	617	642	667	691	716	741		
81	525	550	575	600	625	650	675	700	725	750	.1	3
82	531	557	582	607	633	658	683	709	734	759	.2	5
83	538	564	589	615	640	666	692	717	743	769	.3	8
84	544	570	596	622	648	674	700	726	752	778	.4	10
85	—551	—577	—603	—630	—656	—682	—708	—735	—761	—787	.5	13
86	557	584	610	637	664	690	717	743	770	796	.6	16
87	564	591	618	644	671	698	725	752	779	806	.7	18
88	570	598	625	652	679	706	733	760	788	815	.8	21
89	577	604	632	659	687	714	742	769	797	824	.9	24
90	583	611	639	667	694	722	750	777	806	833		
91	590	618	646	674	702	730	758	786	815	843	.1	3
92	596	625	653	681	710	738	767	795	823	852	.2	6
93	603	631	660	689	718	746	775	804	832	861	.3	9
94	609	638	667	696	725	754	783	812	841	870	.4	12
95	—616	—645	—674	—704	—733	—762	—792	—821	—850	—880	.5	15
96	622	652	681	711	741	770	800	830	859	889	.6	18
97	629	659	689	719	748	778	808	838	868	898	.7	21
98	635	665	696	726	756	786	817	847	877	907	.8	23
99	642	672	703	733	764	794	825	856	886	917	.9	26
100	648	679	710	741	772	802	833	864	895	926		
	21	22	23	24	25	26	27	28	29	30		
	.1	.2	.3	.4	.5	.6	.7	.8	.9	Corrections for tenths in width.		
	1	2	2	3	4	5	5	6	7			

TABLE XI.—*Continued.*

Volumes by the Prismoidal Formula.

Widths.	Heights. 31	32	33	34	35	36	37	38	39	40	Corrections for tenths in height.	
1	10	10	10	10	11	11	11	12	12	12	.1	0
2	19	20	20	21	22	22	23	23	24	25	.2	0
3	29	30	31	31	32	33	34	35	36	37	.3	0
4	38	40	41	42	43	44	46	47	48	49	.4	1
5	48	49	51	52	54	56	57	59	60	62	.5	1
6	57	59	61	63	65	67	68	70	72	74	.6	1
7	67	69	71	73	76	78	80	82	84	86	.7	1
8	77	79	81	84	86	89	91	94	96	97	.8	1
9	86	89	92	94	97	100	103	106	108	111	.9	1
10	96	99	102	105	108	111	114	117	120	123		
11	105	109	112	115	119	122	126	129	132	136	.1	0
12	115	119	122	126	130	133	137	141	144	148	.2	1
13	124	128	132	136	140	144	148	152	156	160	.3	1
14	134	138	143	147	151	156	160	164	169	173	.4	2
15	144	148	153	157	162	167	171	176	181	185	.5	2
16	153	158	163	168	173	178	183	188	193	198	.6	3
17	163	168	173	178	183	189	194	199	205	210	.7	3
18	172	178	183	189	194	200	206	211	217	222	.8	4
19	182	188	194	199	205	211	217	223	229	235	.9	4
20	191	198	204	210	216	222	228	235	241	247		
21	201	207	214	220	227	233	240	246	253	259	.1	1
22	210	217	224	231	238	244	251	258	265	272	.2	2
23	220	227	234	241	248	256	263	270	277	284	.3	2
24	230	237	244	252	259	267	274	281	289	296	.4	3
25	239	247	255	262	270	278	285	293	301	309	.5	4
26	249	257	265	273	281	289	297	305	313	321	.6	5
27	258	267	275	283	292	300	308	317	325	333	.7	5
28	268	277	285	294	302	311	320	328	337	346	.8	6
29	277	286	295	304	313	322	331	340	349	358	.9	7
30	287	296	306	315	324	333	343	352	361	370		
31	297	306	316	325	335	344	354	364	373	383	.1	1
32	306	316	326	336	346	356	365	375	385	395	.2	2
33	316	326	336	346	356	367	377	387	397	407	.3	3
34	325	336	346	357	367	378	388	399	409	420	.4	4
35	335	346	356	367	378	389	400	410	421	432	.5	5
36	344	356	367	378	389	400	411	422	433	444	.6	6
37	354	365	377	388	400	411	423	434	445	457	.7	8
38	364	375	387	399	410	422	434	446	457	469	.8	9
39	373	385	397	409	421	433	445	457	469	481	.9	10
40	383	395	407	420	432	444	457	469	481	494		
41	392	405	418	430	443	456	468	481	494	506	.1	1
42	402	415	428	441	454	467	480	493	506	519	.2	3
43	411	425	438	451	465	478	491	504	518	531	.3	4
44	421	435	448	462	475	489	502	516	530	543	.4	6
45	431	444	458	472	486	500	514	528	542	556	.5	7
46	440	454	469	483	497	511	525	540	554	568	.6	8
47	450	464	479	493	508	522	537	551	566	580	.7	10
48	459	474	489	504	519	533	548	563	578	593	.8	11
49	469	484	499	514	529	544	560	575	590	605	.9	13
50	478	494	509	525	540	556	571	586	602	617		
	31	32	33	34	35	36	37	38	39	40		

	.1	.2	.3	.4	.5	.6	.7	.8	.9	Corrections for tenths in width.
	1	2	3	4	5	6	8	9	10	

TABLE XI.—*Continued.*

Volumes by the Prismoidal Formula.

Widths.	Heights. 31	32	33	34	35	36	37	38	39	40	Corrections for tenths in height.	
51	488	504	519	535	551	567	582	598	614	630	.1	2
52	498	514	530	546	562	578	594	610	626	642	.2	3
53	507	523	540	556	573	589	605	622	638	654	.3	5
54	517	533	550	567	583	600	617	633	650	667	.4	7
55	—526	—543	—560	—577	—594	—611	—628	—645	—662	—679	.5	8
56	586	553	570	588	605	622	640	657	674	691	.6	10
57	545	563	581	598	616	633	651	669	686	704	.7	12
58	555	573	591	609	627	644	662	680	698	716	.8	14
59	565	583	601	619	637	656	674	692	710	728	.9	15
60	574	593	611	630	648	667	685	704	722	741		
61	584	602	621	640	659	678	697	715	734	753	.1	2
62	593	612	631	651	670	689	708	727	746	765	.2	4
63	603	622	642	661	681	700	719	739	758	778	.3	6
64	612	632	652	672	691	711	731	751	770	790	.4	8
65	—622	—642	—662	—682	—702	—722	—742	—762	—782	—802	.5	10
66	631	652	672	693	713	733	754	774	794	815	.6	12
67	641	662	682	703	724	744	765	786	806	827	.7	14
68	651	672	693	714	735	756	777	798	819	840	.8	16
69	660	681	703	724	745	767	788	809	831	852	.9	18
70	670	691	713	735	756	778	799	821	843	864		
71	679	701	723	745	767	789	811	833	855	877	.1	2
72	689	711	733	756	778	800	822	844	867	889	.2	5
73	698	721	744	766	789	811	834	856	879	901	.3	7
74	708	731	754	777	799	822	845	868	891	914	.4	9
75	—718	—741	—764	—787	—810	—833	—856	—880	—903	—926	.5	12
76	727	751	774	798	821	844	868	891	915	938	.6	14
77	737	760	784	808	832	856	879	903	927	951	.7	16
78	746	770	794	819	843	867	891	915	939	963	.8	19
79	756	780	805	829	853	878	902	927	951	975	.9	21
80	765	790	815	840	864	889	914	938	963	988		
81	775	800	825	850	875	900	925	950	975	1000	.1	3
82	785	810	835	860	886	911	936	962	987	1012	.2	5
83	794	820	845	871	897	922	948	973	999	1025	.3	8
84	804	830	856	881	907	933	959	985	1011	1037	.4	10
85	—813	—840	—866	—892	—918	—944	—971	—997	—1023	—1049	.5	13
86	823	849	876	902	929	956	982	1009	1035	1062	.6	16
87	832	859	886	913	940	967	994	1020	1047	1074	.7	18
88	842	869	896	923	951	978	1005	1032	1059	1086	.8	21
89	852	879	906	934	961	989	1016	1044	1071	1098	.9	24
90	861	889	917	944	972	1000	1028	1056	1083	1111		
91	871	900	927	955	983	1011	1039	1067	1095	1123	.1	3
92	880	909	937	965	994	1022	1051	1079	1107	1136	.2	6
93	890	919	947	976	1005	1033	1062	1091	1119	1148	.3	9
94	899	928	957	986	1015	1044	1073	1102	1131	1160	.4	12
95	—909	—938	—968	—997	—1026	—1056	—1085	—1114	—1144	—1173	.5	15
96	919	948	978	1007	1037	1067	1096	1126	1156	1185	.6	18
97	928	958	988	1018	1048	1078	1108	1138	1168	1198	.7	21
98	938	968	998	1028	1059	1089	1119	1149	1180	1210	.8	23
99	947	978	1008	1039	1069	1100	1131	1161	1192	1222	.9	26
100	957	988	1019	1049	1080	1111	1142	1173	1204	1235		
	31	32	33	34	35	36	37	38	39	40		
	.1	.2	.3	.4	.5	.6	.7	.8	.9	Corrections for tenths in width.		
	1	2	3	4	5	6	8	9	10			

TABLE XI.—*Continued.*

VOLUMES BY THE PRISMOIDAL FORMULA.

Widths.	Heights. 41	42	43	44	45	46	47	48	49	50	Corrections for tenths in height.	
1	13	13	13	14	14	14	15	15	15	15	.1	0
2	25	26	27	27	28	28	29	30	30	31	.2	0
3	38	39	40	41	42	43	44	44	45	46	.3	0
4	51	52	53	54	56	57	58	59	60	62	.4	1
5	—63	—65	—66	—68	—69	—71	—73	—74	—76	—77	.5	1
6	76	78	80	81	83	85	87	89	91	93	.6	1
7	89	91	93	95	97	99	102	104	106	108	.7	1
8	101	104	106	109	111	114	116	119	121	123	.8	1
9	114	117	119	122	125	128	131	133	136	139	.9	1
10	127	130	133	136	139	142	145	148	151	154		
11	139	143	146	149	153	156	160	163	166	170	.1	0
12	152	156	159	163	167	170	174	178	181	185	.2	1
13	165	169	173	177	181	185	189	193	197	201	.3	1
14	177	181	186	190	194	199	203	207	212	216	.4	2
15	—190	—194	—199	—204	—208	—213	—218	—222	—227	—231	.5	2
16	203	207	212	217	222	227	232	237	242	247	.6	3
17	215	220	226	231	236	241	247	252	257	262	.7	3
18	228	233	239	244	250	256	261	267	272	278	.8	4
19	240	246	252	258	264	270	276	281	287	293	.9	4
20	253	259	265	272	278	284	290	296	302	309		
21	266	272	279	285	292	298	305	311	318	324	.1	1
22	278	285	292	299	306	312	319	326	333	340	.2	2
23	291	298	305	312	319	327	334	341	348	355	.3	2
24	304	311	319	326	333	341	348	356	363	370	.4	3
25	—316	—324	—332	—340	—347	—355	—363	—370	—378	—386	.5	4
26	329	337	345	353	361	369	377	385	393	401	.6	5
27	342	350	358	367	375	383	392	400	408	417	.7	5
28	354	363	372	380	389	398	406	415	423	432	.8	6
29	367	376	385	394	403	412	421	430	439	448	.9	7
30	380	389	398	407	417	426	435	444	454	463		
31	392	402	411	421	431	410	450	459	469	478	.1	1
32	405	415	425	435	444	454	464	474	484	494	.2	2
33	418	428	438	448	458	469	479	489	499	509	.3	3
34	430	441	451	462	472	483	493	504	514	525	.4	4
35	—443	—454	—465	—475	—486	—497	—508	—519	—529	—540	.5	5
36	456	467	478	489	500	511	522	533	544	556	.6	6
37	468	480	491	502	514	525	537	548	560	571	.7	8
38	481	493	504	516	528	540	551	563	575	586	.8	9
39	494	506	518	530	542	554	566	578	590	602	.9	10
40	506	519	531	543	556	568	580	593	605	617		
41	519	531	544	557	569	582	595	607	620	633	.1	1
42	531	544	557	570	583	596	609	622	635	648	.2	3
43	544	557	571	584	597	610	624	637	650	664	.3	4
44	557	570	584	598	611	625	638	652	665	679	.4	6
45	—569	—583	—597	—611	—625	—639	—653	—667	—681	—694	.5	7
46	582	596	610	625	639	653	667	681	696	710	.6	8
47	595	609	624	638	653	667	682	696	711	725	.7	10
48	607	622	637	652	667	681	696	711	726	741	.8	11
49	620	635	650	665	681	696	710	726	741	756	.9	13
50	633	648	664	679	694	710	725	741	756	772		
	41	42	43	44	45	46	47	48	49	50		
	.1	.2	.3	.4	.5	.6	.7	.8	.9	Corrections for tenths in width.		
	1	3	4	6	7	8	10	11	13			

TABLE XI.—*Continued.*

VOLUMES BY THE PRISMOIDAL FORMULA.

Widths.	Heights. 41	42	43	44	45	46	47	48	49	50	Corrections for tenths in height.	
51	645	661	677	693	708	724	740	756	771	787	.1	2
52	658	674	690	706	722	738	754	770	786	802	.2	3
53	671	687	703	720	736	752	768	785	802	818	.3	5
54	683	700	717	733	750	767	783	800	817	833	.4	7
55	—696	—713	—730	—747	—764	—781	—798	—815	—832	—849	.5	8
56	709	726	743	760	778	795	812	830	847	864	.6	10
57	721	739	756	774	792	809	827	844	862	880	.7	12
58	734	752	770	788	806	823	841	859	877	895	.8	14
59	747	765	783	801	819	833	856	874	892	910	.9	15
60	759	778	796	815	833	852	870	889	907	926		
61	772	791	810	828	847	866	885	904	923	941	.1	2
62	785	804	823	842	861	880	899	919	938	957	.2	4
63	797	817	836	856	875	894	914	933	953	972	.3	6
64	810	830	849	869	889	909	928	948	968	988	.4	8
65	—823	—843	—863	—883	—903	—923	—943	—963	—983	—1003	.5	10
66	835	856	876	896	917	937	957	978	998	1019	.6	12
67	848	869	889	910	931	951	972	993	1013	1034	.7	14
68	860	881	902	923	944	965	986	1007	1028	1049	.8	16
69	873	894	916	937	958	980	1001	1022	1044	1065	.9	18
70	886	907	929	951	972	994	1015	1037	1059	1080		
71	898	920	942	964	986	1008	1030	1052	1074	1096	.1	2
72	911	933	956	978	1000	1022	1044	1067	1089	1111	.2	5
73	924	946	969	991	1014	1036	1059	1081	1104	1127	.3	7
74	936	959	982	1005	1028	1051	1073	1096	1119	1142	.4	9
75	—949	—972	—995	—1019	—1042	—1065	—1088	—1111	—1134	—1157	.5	12
76	962	985	1009	1032	1056	1079	1102	1126	1149	1173	.6	14
77	974	998	1022	1046	1069	1093	1117	1141	1165	1188	.7	16
78	987	1011	1035	1059	1083	1107	1131	1156	1180	1204	.8	19
79	1000	1024	1048	1073	1097	1122	1146	1170	1195	1219	.9	21
80	1012	1037	1062	1086	1111	1136	1160	1185	1210	1235		
81	1025	1050	1075	1100	1125	1150	1175	1200	1225	1250	.1	3
82	1038	1063	1088	1114	1139	1164	1190	1215	1240	1265	.2	5
83	1050	1076	1102	1127	1153	1178	1204	1230	1255	1281	.3	8
84	1063	1089	1115	1141	1167	1193	1219	1244	1270	1296	.4	10
85	—1076	—1102	—1128	—1154	—1181	—1207	—1233	—1259	—1285	—1312	.5	13
86	1088	1115	1141	1168	1194	1221	1248	1274	1301	1327	.6	16
87	1101	1128	1155	1181	1208	1235	1262	1289	1316	1343	.7	18
88	1114	1141	1168	1195	1222	1249	1277	1304	1331	1358	.8	21
89	1126	1154	1181	1209	1236	1264	1291	1319	1346	1373	.9	24
90	1139	1167	1194	1222	1250	1278	1306	1333	1361	1389		
91	1152	1180	1208	1236	1264	1292	1320	1348	1376	1404	.1	3
92	1164	1193	1221	1249	1278	1306	1335	1363	1391	1420	.2	6
93	1177	1206	1234	1263	1292	1320	1349	1378	1406	1435	.3	9
94	1190	1219	1248	1277	1306	1335	1364	1393	1422	1451	.4	12
95	—1202	—1231	—1261	—1290	—1319	—1349	—1378	—1407	—1437	—1466	.5	15
96	1215	1244	1274	1304	1333	1363	1393	1422	1452	1481	.6	18
97	1227	1257	1287	1317	1347	1377	1407	1437	1467	1497	.7	21
98	1240	1270	1301	1331	1361	1391	1422	1452	1482	1512	.8	23
99	1253	1283	1314	1344	1375	1406	1436	1467	1497	1528	.9	26
100	1265	1296	1327	1358	1389	1420	1451	1481	1512	1543		
	41	42	43	44	45	46	47	48	49	50		
	.1	.2	.3	.4	.5	.6	.7	.8	.9	Corrections for tenths in width.		
	1	3	4	6	7	8	10	11	13			

TABLE XII.—AZIMUTHS OF POLARIS

THE STAR AND THE AZIMUTH are W. of N. when the hour angle is *less*
THE ARGUMENT is the star's hour angle (or 23h. 56min.
TO FIND THE TRUE MERIDIAN the azimuth must be laid off to the *east* when the

Hours.	1892.	1894.	1896.	1898.	1900.	Azimuths for latitude— 30°	32°	34°	36°	38°	40°	42°	44°	46°	48°	50°
h.	m.	m.	m.	m.	m.	° ′	° ′	° ′	° ′	° ′	° ′	° ′	° ′	° ′	° ′	° ′
0	4	4	4	4	4	0 2	0 2	0 2	0 2	0 2	0 2	0 2	0 2	0 2	0 2	0 2
	8	8	8	8.	8.	3	3	3	3	3	4	4	4	4	4	4
	12	12	12.	12.	12.	5	5	5	5	5	5	6	6	6	6	6
	16	16.	16.	16.	16.	6	6	7	7	7	7	7	8	8	8	8
	20	20.	20.	20.	21	8	8	8	8	9	9	9	9	10	10	11
	24	24.	24.	24.	25	9	10	10	10	10	11	11	11	12	12	13
	28	28.	28.	29	29	11	11	11	12	12	12	13	13	14	14	15
	32	32.	32.	33	33.	12	13	13	13	14	14	15	15	16	16	17
	36	36.	37	37	37.	14	14	15	15	15	16	16	17	18	18	19
	40	40.	41	41.	41.	15	16	16	17	17	18	18	19	20	20	21
	44.	44.	45	45.	46	17	17	18	18	19	19	20	21	21	23	23
	48.	49	49	49.	50	19	19	19	20	21	22	22	23	23	24	25
	52.	53	53.	53.	54	20	21	21	22	22	23	24	24	25	26	27
0	56.	57	57.	58	58.	22	22	23	23	24	25	25	26	27	28	29
1	0.	1	1.	2	2.	23	24	24	25	26	26	27	28	29	30	32
	5.	6	6.	7	7.	25	26	26	27	28	28	29	30	31	33	34
	10.	11	11.	12.	13	27	27	28	29	30	31	32	33	34	35	37
	15.	16	17	17.	18	29	29	30	31	32	33	34	35	36	38	39
	20.	21.	22	22.	23.	31	31	32	33	34	35	36	37	38	40	42
	25.	26.	27	28.	28.	32	33	34	35	36	37	38	39	41	42	44
	31	31.	32.	33.	31	34	35	36	37	38	39	40	42	43	45	47
	35.	36.	37.	38	39	36	37	38	39	40	41	42	44	45	47	49
	40.	41.	42.	43.	44.	38	39	40	41	42	43	44	46	47	49	51
	45.	46.	47.	48.	49.	39	40	41	42	44	45	46	48	50	52	54
	50.	52	53	54	55	41	42	43	44	46	47	48	50	52	54	56
	55.	57	58	59	0	43	44	45	46	47	49	50	52	54	56	0 59
1	1	2	3	4.	5.	45	46	47	48	49	51	52	54	56	0 58	1 1
2	6	7.	8.	9.	10.	46	47	49	50	51	53	54	56	0 58	1 1	3
	11	12	13.	14.	16	48	49	50	51	53	54	56	0 58	1 0	3	5
	16	17.	18.	20	21.	50	51	52	53	55	56	0 58	1 0	2	5	8
	21	22.	24	25	26.	51	52	54	55	57	0 58	1 0	2	4	7	10
	26	27.	29	30.	32	53	54	55	57	0 58	1 0	2	4	6	9	12
	31	32.	34.	35.	37	54	55	57	0 58	1 0	2	4	6	8	11	14
	36	38	39.	41	42.	56	57	0 58	1 0	2	3	6	8	10	13	16
	41	43	44.	46.	48	57	0 59	1 0	2	3	5	7	9	12	15	18
	46	48	49.	51.	53	0 59	1 0	2	3	5	7	9	11	14	17	20
	51	53.	55	57	58.	1 0	2	3	5	7	8	11	13	16	19	22
	56.	58.	0	2	4	2	3	5	6	8	10	12	15	17	20	24
2	1.	3.	5.	7.	9.	3	4	6	8	10	12	14	16	19	22	26
3	7.	10	12	14	16	5	6	8	10	12	13	16	18	21	24	28
	13.	16	18	20.	23	6	8	9	11	13	15	18	20	23	27	30
	19.	23	24.	27	29.	8	9	11	13	15	17	19	22	25	28	32
	26	29	31	33.	36	9	11	13	14	16	19	21	24	27	30	34
	32	35	37.	40	43	11	13	14	16	18	20	23	25	29	32	36
	39	42	45	48	51	12	14	16	17	20	22	25	27	31	34	38
	46.	49.	52.	55.	50	14	15	17	19	21	24	26	29	32	36	40
	53.	57	0	3.	7	15	17	19	21	23	25	28	31	34	38	42
3	2.	6.	10	13.	17.	17	19	21	23	25	27	30	33	36	40	44
4	13	17.	21.	25.	30	19	21	23	24	27	29	32	35	39	43	47
	23.	28.	33	38	43	20	22	24	26	29	31	34	37	41	45	49
	34	40	45	50.	57.	22	24	26	28	30	33	36	39	42	47	51
	50	57.	4.	12.	23	24	26	28	30	33	35	38	41	45	49	54
4	7	17	29	50.	....	26	27	30	32	34	37	40	43	47	51	56
5	32	....	....	....	...	27	29	31	33	36	39	42	45	49	53	58
5	....	....	....	....	....	1 29	1 30	1 32	1 35	1 37	1 40	1 43	1 47	1 50	1 55	1 59

Date. 1893.	
Jan.	1
	15
Feb.	1
	15
Mar.	1
	15
Apr.	1
	15
May	1
	15
June	1
	15
July	1
	15
Aug.	1
	15
Sept.	1
	15
Oct.	1
	15
Nov.	1
	15
Dec.	1
	15

Tabular
Days.
1
2
3
4
5
6
7
8
9
10
11
12
13
14
15
16

FOR ALL HOUR ANGLES. § 381A.

than 11h 58m and E. of N. when the hour angle is *greater* than 11h 58m.
minus the star's hour angle), for the years given.
hour angle is *less* than 11h 58m, and to the *west* when it is *greater* than 11h 58m.

Time of upper Culmination after mean noon.	Hours.	1892.	1894.	1896.	1898.	1900.	Azimuths for latitude— 30°	32°	34°	36°	38°	40°	42°	44°	46°	48°	50°
h. m.	h.	m.	m.	m.	m.	m.	° ′	° ′	° ′	° ′	° ′	° ′	° ′	° ′	° ′	° ′	° ′
6 32.3	11	54	54	54	54	54	0 1	0 1	0 1	0 2	0 2	0 2	0 2	0 2	0 2	0 2	0 2
		50	50	50	50	50	3	3	3	3	3	3	3	4	4	4	4
5 37.0		46	46	46	45.	45.	5	5	5	5	5	5	5	5	6	6	6
4 29.9		42	42	41.	41.	41.	6	6	6	6	7	7	7	7	8	8	8
		38	37.	37.	37.	37.	8	8	8	8	8	8	9	9	9	10	10
3 34.6		34	33.	33.	33.	33	9	9	9	10	10	10	11	11	11	12	12
2 39.4		30	29.	29.	29	29	11	11	11	11	12	12	12	13	13	14	14
1 44.3		26	25.	25.	25	25	12	12	13	13	13	14	14	14	15	15	16
		22	21.	21	21	21	14	14	14	15	15	15	16	16	17	17	18
0 37.3		18	17.	17	17	16.	15	15	16	16	17	17	18	18	19	19	20
23 38.4		14	13.	13	12.	12.	17	17	17	18	18	19	19	20	21	21	22
		10	9.	9	8.	8	18	18	19	19	20	20	21	22	22	23	25
22 35.5		6	5.	5	4.	4	20	20	20	21	22	22	23	23	24	25	26
21 40.6	11	2	1.	1	0.	0	21	21	22	23	23	24	24	25	26	27	28
20 34.0	10	57.	57	56.	56.	55.	23	23	24	24	25	25	26	27	28	29	30
		52.	52	51.	51	50.	24	25	25	26	27	27	28	29	30	31	32
19 39.1		47.	47	46.	46	45	26	27	27	28	29	29	30	31	32	34	35
18 36.5		42.	42.	41.	40.	40	28	29	29	30	31	31	32	33	35	36	37
		37.	37	36	35.	35	30	30	31	22	33	34	35	36	37	38	40
17 41.6		32.	32	31	30.	29.	32	32	33	34	35	36	37	38	39	41	42
16 35.1		27.	26.	26	25	24.	33	34	35	36	37	38	39	40	41	43	44
15 40.2		22.	21.	21	20	19	35	36	37	37	39	40	41	42	43	45	47
		17.	16.	15.	15	14	37	38	39	39	40	41	43	44	46	47	49
14 33.6		12.	11.	10.	9.	8.	39	39	40	41	42	43	45	46	48	49	51
13 38.7		7.	6.	5.	4.	3.	40	41	42	43	44	45	47	48	50	52	54
	10	2.	1.	0	59	58	42	43	44	45	46	47	49	50	52	54	56
12 35.9	9	57.	56.	55.	54	53	44	45	46	47	48	49	51	52	54	56	0 58
11 40.8		52.	51	50.	49	47.	45	46	47	48	50	51	53	54	56	0 58	1 0
10 34.0		47.	46	44.	43.	42	47	48	49	50	51	53	54	56	0 58	1 0	2
		42	40.	39.	38.	37	49	50	51	52	53	55	56	0 58	1 0	2	5
9 38.9		37	35.	34.	33	31.	50	51	52	53	55	56	0 58	1 0	2	4	7
8 35.8		32	30.	29	28	26.	51	53	54	55	57	0 58	1 0	2	4	6	9
		27	25.	24	22.	21	53	54	56	57	0 58	1 0	2	4	6	8	11
7 40.6		22	20.	19	17.	15.	55	56	57	0 58	1 0	2	4	5	8	10	13
		17	15	13.	12	10	56	57	0 59	1 0	2	3	5	7	10	12	15
		12	10	8.	6.	5	58	0 59	1 0	2	3	5	7	9	11	14	17
difference.		7	5	3	1.	59.	0 59	1 0	2	3	5	7	9	11	13	16	19
	9	2	59.	58	56	54	1 0	2	3	5	6	8	10	12	15	18	21
	8	56.	54.	52.	51	49	2	3	5	6	8	10	12	14	17	20	23
m.		50.	48.	40.	44.	42.	3	5	6	8	10	12	14	16	10	22	25
8.9		44.	42.	40	38	35.	5	6	8	9	11	13	16	18	21	24	27
7.9		38.	36	33.	31.	29	7	8	10	11	13	15	17	20	22	26	29
11.6		32.	29.	27	25	23.	8	9	11	13	15	17	19	22	24	28	31
15.7		26	23	21	17.	15.	10	11	13	14	16	18	21	23	26	20	33
19.6		19	16	13.	10.	7.	11	13	14	16	18	20	23	25	28	31	35
23.6		12	8.	5.	2.	59.	13	14	16	18	20	22	25	27	30	33	37
27.5	8	5	1	58	55	51.	14	16	18	19	21	24	26	29	32	35	39
31.4	7	55	51.	48	44.	40.	16	18	19	21	24	26	28	31	34	38	42
35.4		45	40.	37	33	28.	18	19	21	23	26	28	31	33	37	40	44
39.8		34.	29.	25	20.	15.	20	21	23	25	27	30	33	35	39	43	47
43.2		24	18.	13.	7.	1.	21	23	25	27	29	32	35	37	41	45	49
47.2	7	8	1	54	45.	35.	23	25	27	29	32	34	37	40	43	47	52
51.1																	
55.0	6	51.	41	30	11	....	25	27	29	31	34	36	39	42	46	50	54
58.9	6	26	....	....	....	....	27	29	31	33	36	38	41	44	48	52	57
62.9	5	....	....	....	....	....	1 29	1 30	1 32	1 35	1 37	1 40	1 43	1 47	1 50	1 55	1 59

www.ingramcontent.com/pod-product-compliance
Lightning Source LLC
Chambersburg PA
CBHW021943190326
41519CB00009B/1126
* 9 7 8 3 3 3 7 0 2 4 5 7 4 *